Lecture Notes in Statistics 147

Edited by P. Bickel, P. Diggle, S. Fienberg, K. Krickeberg,
I. Olkin, N. Wermuth, S. Zeger

W0231951

Springer-Science+Business Media, LLC

Jürgen Franke
Wolfgang Härdle
Gerhard Stahl

Editors

Measuring Risk in
Complex Stochastic Systems

 Springer

Library of Congress Cataloging-in-Publication Data

Measuring risk in complex stochastic systems / Jürgen Franke, Wolfgang Härdle,
Gerhard Stahl, editors.
   p. cm. -- (Lecture notes in statistics ; 147)
   Includes bibliographical references and index.
   ISBN 978-0-387-98996-9   ISBN 978-1-4612-1214-0 (eBook)
   DOI 10.1007/978-1-4612-1214-0
   1. Risk management--Mathematical models. 2. Investments--Mathematical models. 3.
Finance--Mathematical models. 4. Asset-liability management. I. Franke, Jürgen. II.
Härdle, Wolfgang. III. Stahl, Gerhard. IV. Lecture notes in statistics (Springer-Verlag) ;
147.

HD61 .M43 2000
658.15'5--dc21

00-026569

Printed on acid-free paper.

© 2000 Springer Science+Business Media New York
Originally published by Springer-Verlag New York in 2000

Camera ready copy provided by the authors

9 8 7 6 5 4 3 2 1

ISBN 978-0-387-98996-9

Preface

Complex dynamic processes of life and sciences generate risks that have to be taken. The need for clear and distinctive definitions of different kinds of risks, adequate methods and parsimonious models is obvious. The identification of important risk factors and the quantification of risk stemming from an interplay between many risk factors is a prerequisite for mastering the challenges of risk perception, analysis and management successfully. The increasing complexity of stochastic systems, especially in finance, have catalysed the use of advanced statistical methods for these tasks.

The methodological approach to solving risk management tasks may, however, be undertaken from many different angles. A financial institution may focus on the risk created by the use of options and other derivatives in global financial processing, an auditor will try to evaluate internal risk management models in detail, a mathematician may be interested in analysing the involved nonlinearities or concentrate on extreme and rare events of a complex stochastic system, whereas a statistician may be interested in model and variable selection, practical implementations and parsimonious modelling. An economist may think about the possible impact of risk management tools in the framework of efficient regulation of financial markets or efficient allocation of capital.

This book gives a diversified portfolio of these scenarios. We first present a set of papers on credit risk management, and then focus on extreme value analysis. The Value at Risk (VaR) concept is discussed in the next block of papers, followed by several articles on change points. The papers were presented during a conference on Measuring Risk in Complex Stochastic Systems that took place in Berlin on September 25th - 30th 1999. The conference was organised within the Seminar Berlin-Paris, Seminaire Paris-Berlin.

The paper by Lehrbass considers country risk within a no-arbitrage model and combines it with the extended Vasicek term structure model and applies the developed theory to DEM- Eurobonds. Kiesel, Perraudin and Taylor construct a model free volatility estimator to investigate the long horizon volatility of various short term interest rates. Hanousek investigates the failing of Czech banks during the early nineties. Müller and Rnz apply a Generalized Partial Linear Model to evaluating credit risk based on a credit scoring data set from a French bank. Overbeck considers the problem of capital allocation in the framework of credit risk and loan portfolios.

The analysis of extreme values starts with a paper by Novak, who considers confidence intervals for tail index estimators. Robert presents a novel approach to extreme value calculation on state of the art α-ARCH models. Kleinow and Thomas show how in a client/server architecture the computation of extreme value parameters may be undertaken with the help of WWW browsers and an XploRe Quantlet Server.

The VaR section starts with Cumperayot, Danielsson and deVries who discuss basic questions of VaR modelling and focus in particular on economic justifications for external and internal risk management procedures and put into question the rationale behind VaR.

Slaby and Kokoschka deal with with change-points. Slaby considers methods based on ranks in an iid framework to detect shifts in location, whereas Kokoszka reviews CUSUM-type esting and estimating procedures for the change-point problem in ARCH models.

Huschens and Kim concentrate on the stylised fact of heavy tailed marginal distributions for financial returns time series. They model the distributions by the family of α-stable laws and consider the consequences for β values in the often applied CAPM framework. Breckling, Eberlein and Kokic introduce the generalised hyperbolic model to calculate the VaR for market and credit risk. Härdle and Stahl consider the backtesting based on shortfall risk and discuss the use of exponential weights. Sylla and Villa apply a PCA to the implied volatility surface in order to determine the nature of the vola factors.

We gratefully acknowledge the support of the Deutsche Forschungsgemeinschaft, SFB 373 Quantification und Simulation Ökonomischer Prozesse, Weierstra Institut für Angewandte Analysis und Stochastik, Deutsche Bank, WestLB, BHF-Bank, Arthur Andersen, SachsenLB, and MD*Tech.

The local organization was smoothly run by Jörg Polzehl and Vladimir Spokoiny. Without the help of Anja Bardeleben, Torsten Kleinow, Heiko Lehmann, Marlene Müller, Sibylle Schmerbach, Beate Siegler, Katrin Westphal this event would not have been possible.

J. Franke, W. Härdle and G. Stahl

January 2000, Kaiserslautern and Berlin

Contributors

Jens Breckling Insiders GmbH Wissensbasierte Systeme, Wilh.-Th.--Römheld-Str. 32, 55130 Mainz, Germany

Phornchanok J. Cumperayot Tinbergen Institute, Erasmus University Rotterdam

Jon Danielsson London School of Economics

Casper G. de Vries Erasmus University Rotterdam and Tinbergen Institute

Ernst Eberlein Institut für Mathematische Stochastik, Universität Freiburg, Eckerstaße 1, 79104 Freiburg im Breisgau, Germany

Wolfgang Härdle Humboldt-Universität zu Berlin, Dept. of Economics, Spandauer Str. 1, 10178 Berlin

Jan Hanousek CERGE-EI, Prague

Stefan Huschens Technical University Dresden, Dept. of Economics

Bjorn N. Jorgensen Harvard Business School

Rüdiger Kiesel School of Economics, Mathematics and Statistics, Birkbeck College, University of London, 7-15 Gresse St., London W1P 2LL, UK

Jeong-Ryeol Kim Technical University Dresden, Dept. of Economics

Torsten Kleinow Humboldt-Universität zu Berlin, Dept. of Economics, Spandauer Str. 1, 10178 Berlin

Philip Kokic Insiders GmbH Wissensbasierte Systeme, Wilh.-Th.-Römheld-Str. 32, 55130 Mainz, Germany

Piotr Kokoszka The University of Liverpool and Vilnius University Institute of Mathematics and Informatics

Frank Lehrbass 01-616 GB Zentrales Kreditmanagement, Portfoliosteuerung, WestLB

Marlene Müller Humboldt-Universität zu Berlin, Dept. of Economics, Spandauer Str. 1, 10178 Berlin

Sergei Y. Novak EURANDOM PO Box 513, Eindhoven 5600 MB, Netherlands

Ludger Overbeck Deutsche Bank AG, Group Market Risk Management, Methodology & Policy/CR, 60262 Frankfurt

William Perraudin Birkbeck College, Bank of England and CEPR

Christian Robert Centre de Recherche en Economie et Statistique (CREST), Laboratoire de Finance Assurance, Timbre J320 - 15, Bb G. Peri, 92245 MALAKOFF, FRANCE

Bernd Rönz Humboldt-Universität zu Berlin, Dept. of Economics, Spandauer Str. 1, 10178 Berlin

Aleš Slabý Charles University Prague, Czech Republic

Gerhard Stahl Bundesaufsichtsamt für das Kreditwesen, Berlin

Alpha Sylla ENSAI-Rennes, Campus de Ker-Lan, 35170 Bruz, France.

Alex Taylor School of Economics, Mathematics and Statistics, Birkbeck College, University of London, 7-15 Gresse St., London W1P 2LL, UK

Michael Thomas Fachbereich Mathematik, Universität-Gesamthochschule Siegen

Christophe Villa University of Rennes 1, IGR and CREREG, 11 rue jean Mac, 35019 Rennes cedex, France.

Contents

1 Allocation of Economic Capital in loan portfolios

Ludger Overbeck

1.1 Introduction

Since the seminal research of Markowitz (1952) and Sharpe (1964) capital allocation within portfolios is based on the variance/covariance analysis. Even the introduction of Value-at-Risk in order to measure risk more accurately than in terms of standard deviation, did not chance the calculation of a risk contribution of single asset in the portfolio or its contributory capital as a multiple of the asset's β with the portfolio. This approach is based on the assumption that asset returns are normally distributed. Under this assumption, the capital of a portfolio, usually defined as a quantile of the distribution of changes of the portfolio value, is a multiple of the standard deviation of the portfolio. Since the βs yield a nice decomposition of the portfolio standard deviation and exhibit the interpretation as an infinitesimal marginal risk contribution (or more mathematically as a partial derivative of the portfolio standard deviation with respect to an increase of the weight of an asset in the portfolio), these useful properties also hold for the quantile, i.e. for the capital.

In the case of the normal distributed assets in the portfolio, the though defined capital allocation rule also coincides with the capital allocation based on marginal economic capital, i.e. the capital difference between the portfolio with and without the single asset to which we want to allocate capital. Additionally it is equivalent to the expected loss in the single asset conditional on the event that the loss for the whole portfolio exceeds a quantile of the loss distribution.

The purpose of the paper is to present and analyse these three capital allocation rules, i.e. the one based on conditional expectation, the one on marginal economic capital and the classical one based on covariances, in the context of a loan portfolio. The only method that gives analytic solutions of the (relative) allocation rule is the classical one based on covariances. All others have to be analysed by a Monte-Carlo-Simulation for real world portfolios. There is of course a possibility to quantify the other two approaches for highly uniformed and standardized portfolios. On the other hand in some situations also the calculation of the βs might be quicker in a Monte-Carlo-Simulation.

1.2   Credit portfolios

Let us consider a portfolio of transactions with m counterparties. The time horizon at which the loss distribution is to be determined is fixed, namely 1 year. The random variable portfolio loss can than be written as

$$L \;=\; \sum_{k=1}^{m} L_k, \tag{1.1}$$

where L_k is the loss associated with transaction k. There are now different models discussed in the literature and some of them are implemented by banks and software firms. From the growing literature the papers Baestaens and van den Bergh (1997), *Credit Metrics* (1997), Risk (1997), Artzner, Dealban, Eber and Heath (1997a), Kealhofer (1995), Overbeck and Stahl (1997), Schmid (1997), Vasicek (1997) and Wilson (1997) may be consulted in a first attempt.

In the simplest model (pure default mode)

$$L_k \;=\; l_k \mathbf{1}_{D_k}, \tag{1.2}$$

where D_k is the default event and l_k is the exposure amount, which is assumed to be known with certainty. More eloborate models (like *Credit Metrics* (1997)) assume

$$L_k \;=\; \sum_{r=AAA}^{D} l_{r,k} \mathbf{1}_{D_{r,k}}, \tag{1.3}$$

where $D_{r,k}$ the event that counterparty k is in rating class r and $l_{r,k}$ is the loss associated with the migration of k to rating r. The loss amount is usually deterministic given the total amount of exposure and the given migration, i.e. $l_{r,k}$ is a function of r, the exposure and the present rating of k. The straight asset value model Merton (1974), e.g. implemented by Kealhofer and Crosbie (1997), assumes

$$L_k \;=\; L(k, A_1(k)), \tag{1.4}$$

where $A(k)$ is the stochastic process governing the asset value process of counterparty k. In the default mode only model

$$L(k, A_1(k)) = l_k \mathbf{1}_{\{A_1(k) < C_k\}}, \qquad (1.5)$$

where C_k is the default boundary. We will basically consider the last approach, but similar results also hold for more general models like (1.1).

1.2.1 Ability to Pay Process

In the model descriptions (1.5) and (1.4) the process driving default is usually addressed as the asset-value process. This originated in the seminal paper by Merton (1974). The main area of application is the default analysis of firms with stock exchange traded equities. However a straightforward abstraction leads to the formulation of an "Ability to pay process". If this process falls below under a certain threshold then default occurs. However in general the modeling of the ability to pay of a given customer is difficult.

Nevertheless let us assume we have m customer with exposure $l_k, k = 1, .., m$ and ability to pay process

$$dA_t(i) = \mu_i A_t(i)dt + \sigma_i A_t(i)dZ_t(i). \qquad (1.6)$$

Here $Z_t = (Z_t(1), .., Z_t(m))$ is a standard multivariate Brownian motion with covariance matrix equal to correlation matrix $R = (\rho_{ij})$. If now the threshold C_k were known, the distribution of L would be specify. Since the parameters of the ability to pay process are difficult to access, we take another route here. We just assume that the default probability for each single customer and the correlation matrix R is known. Default probabilities can be calibrated from the spread in the market or from historical default data provided by rating agencies or by internal ratings. The correlation may be derived from equity indices as proposed in the *Credit Metrics* (1997) model. This two sets of parameters are sufficient since

$$
\begin{aligned}
L &= \sum_{k=1}^{m} l_k \mathbf{1}_{\{A_1(k) < C_k\}} \\
&= \sum_{k}^{m} l_k \mathbf{1}_{\{A_0(k) \exp\{(\mu_k - \frac{1}{2}\sigma_k) + \sigma_k Z_1\} < C_i\}} \\
&= \sum_{k}^{m} l_k \mathbf{1}_{\{Z_1(k) < \frac{\log C_i - \log A_0(k) - \mu_k - \frac{1}{2}\sigma_k}{\sigma_k}\}} \\
&= \sum_{k}^{m} l_k \mathbf{1}_{\{Z_1(k) < \Phi^{-1}(p_k)\}},
\end{aligned}
\tag{1.7}
$$

where p_k is the default probability of counterparty k and Φ is the distribution function of the standard normal distribution. Hence the distribution of the vector Z_1, i.e. the correlation R and the default probabilities specify the loss distribution entirely. Remember that we assumed the l_k to be non-random.

1.2.2   Loss distribution

There are attempts to give an analytic approximation to the distribution of L. If all $p_i = p_0$ and all correlation are the same and all exposures are equal then a straight forward application of some limit theorems like LLN,CLT,Poisson law give different reasonable approximations for large m. This is for example discussed in Fingers (1999).

Since the analyzed capital allocation rules require all but one Monte-Carlo-Simulation we also simulate the loss distribution itself. The empirical distribution

$$
\frac{1}{N} \sum_{i=1}^{N} \mathbf{1}_{[0,x]} \left(\sum_{k=1}^{m} l_k \mathbf{1}_{\{Z_1^i(k) < \Phi^{-1}(p_k)\}} \right),
\tag{1.8}
$$

where N is the number of simulation and the vector Z_1^i is the i-th realization of a multivariate normal distribution with correlation matrix R, serves as an approximation of the true loss distribution. A typical histogram of a simulated loss distribution is shown below in Figure 1. It shows the 10% largest losses in the simulation of the portfolio described in Section 7 below.

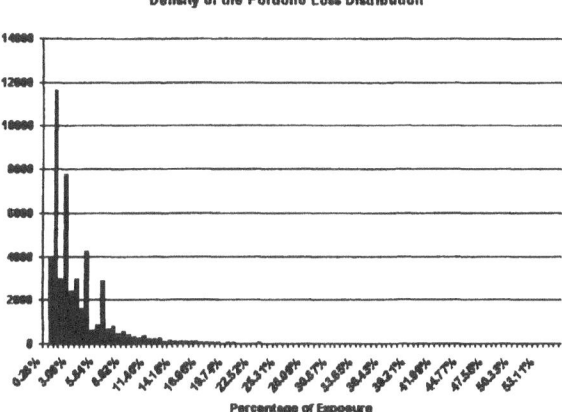

Figure 1.1: Histogram of a simulated loss distribution

1.3 Economic Capital

The nowadays widespread definition of economic capital for a portfolio of financial instruments uses the notion of the quantile of the loss distribution. Economic capital, based on a confidence of $\alpha\%$, $\text{EC}(\alpha)$ is set to the α-quantile of the loss distribution minus the expected value of the loss distribution, more precisely

$$\text{EC}(\alpha) \;=\; q_\alpha(L) - E[L], \text{with} \tag{1.9}$$

$$q_\alpha(L) \;=\; \inf\{y|P[L > y] > 1 - \frac{\alpha}{100}\}. \tag{1.10}$$

From a risk management point of view, holding the amount $\text{EC}(99.98)$ as cushion against the portfolio defining L means that in average in 4999 out of 5000 years the capital would cover all losses. This approach towards economic capital resembles an "all or nothing" rule. In particular in "bad" times, when 1 out of this 5000 events happens, the capital does not cushion the losses. If L is based on the whole balance sheet of the bank and there is no additional capital, the bank would be in default itself. An alternative capital definition tries also to think about "bad times" a little more optimistic. Let "bad times" be specified by the event, that the loss is bigger than a given amount K and let economic capital be defined by

$$\text{EC}_K \;=\; E[L|L > K]. \tag{1.11}$$

This economic capital is in average also enough to cushion even losses in bad times. This approach motives also our definition of contributory capital based on coherent risk measures. This capital definition is analyzed in detail by Artzner, Dealban, Eber and Heath (1997b). They also show that EC_K is coherent if K is a quantile of L. Coherency requires a risk measure to satisfy a set of axiom, or first principles, that a reasonable risk measure should obey. It is also shown that the risk measure defined in terms of quantiles are not coherent in general.

1.3.1   Capital allocation

Once there is an agreement about the definition and the amount of capital EC, it is often necessary to allocate it throughout the portfolio. We therefore look for a contributory economic capital γ_k for each $k = 1, .., m$ such that

$$\sum_{k=1}^{m} \gamma_k \ = \ \text{EC}. \tag{1.12}$$

1.4   Capital allocation based on Var/Covar

The classical portfolio theory provides a rule for the allocation of contributory economic capital that is based on the decomposition of the standard deviation of the loss distribution. These contributions to the standard deviation are called Risk Contributions β_i. By construction of the random variable L in (1.5) we have

$$\begin{aligned}
\sigma^2(L) \ &= \ \sum_{k=1}^{m} l_k^2 \sigma^2 \big(1_{\{Z_1(k) < \Phi^{-1}(p_k)\}}\big) \tag{1.13} \\
&\quad +2 \sum_{i=1}^{m} \sum_{j=i+1}^{m} l_i l_j \text{cov}\big(1_{\{Z_1(i) < \Phi^{-1}(p_i)\}}, 1_{\{Z_1(j) < \Phi^{-1}(p_j)\}}\big) \\
&= \ \sum_{k=1}^{m} l_k p_k (1 - p_k) \\
&\quad +2 \sum_{i=1}^{m} \sum_{j=i+1}^{m} l_i l_j P[Z_1(i) < \Phi^{-1}(p_i), Z_1(j) < \Phi^{-1}(p_j)].
\end{aligned}$$

Moreover equation (1.13) yields immediately that with

$$\beta_i := \frac{1}{\sigma(L)}\left((l_i p_i (1-p_i) \right.$$

$$\left. + \sum_{j=1,j\neq i}^{m} l_j \left(P[Z_1(i) < \Phi^{-1}(p_i), Z_1(j) < \Phi^{-1}(p_j)] - p_i p_j \right) \right),$$

we obtain

$$\sum_{k=1}^{m} l_k \beta_k \quad = \sigma(L). \tag{1.14}$$

Also $P[Z_1(i) < \Phi^{-1}(p_i), Z_1(j) < \Phi^{-1}(p_j)]$ is easily calculated as the integral of a two dimensional normalized normal distribution with correlation r_{ij}.

For the family of normally distributed random variable the difference between a quantile and the mean is simple a multiple of the standard deviation. This is the historical reason why risk managers like to think in multiples of the standard deviation, or the volatility to quantify risk.

This approach is also inherited to non-normal distribution, since also in credit risk management the contributory economic capital is often defined by

$$\gamma_k = \beta_k \cdot \frac{EC(\alpha)}{\sigma_L}.$$

The β_k can also be viewed as infinitesimal small marginal risk or more mathematically

$$\beta_i = \frac{\partial \sigma(L)}{\partial l_i}.$$

If the portfolio L were a sum of normal distributed random variables weighted by l_k we would also have

$$\gamma_i = \frac{\partial}{\partial l_i} EC(\alpha)$$

$$= \frac{EC(\alpha)}{\sigma(L)} \cdot \frac{\partial}{\partial l_i} \sigma_i$$

$$= \frac{EC(\alpha)}{\sigma(L)} \cdot \beta_i$$

as intended. This interpretation breaks down if L is not a linear function of a multivariate normal distributed random vector. We therefore analyze marginal economic capital in the very definition in the following section.

1.5   Allocation of marginal capital

Marginal capital for a given counterparty j, $MEC_j(\alpha)$ is defined to be the difference between the economic capital of the whole portfolio and the economic capital of the portfolio without the transaction:

$$MEC_j(\alpha) \;\; = \;\; \text{EC}(\alpha, L) - \text{EC}(\alpha, L - l_j \mathbf{1}_{\{Z_1(j) < \Phi^{-1}(p_j)\}}).$$

Since the sum of the MECs does not add up to $EC(\alpha)$ we either define the economic capital to be the sum of the MECs or allocate the contributory economic capital proportional to the marginal capital, i.e.

$$CEC_j(II) \;\; = \;\; \text{MEC}_j \frac{\text{EC}(\alpha)}{\sum_{k=1}^{m} \text{MEC}_k}. \tag{1.15}$$

Since the sum of the CECs has no significant economic interpretation we define the capital allocation rule based on marginal capital by (1.15).

1.6   Contributory capital based on coherent risk measures

There are doubt whether the definition of capital in terms of quantiles is useful. In Artzner et al. (1997b) a different route is taken. They go back to "first principles" and ask which are the basic features a risk measure should have. Measures satisfying these axioms are called coherent. They are already used in insurance mathematics and extreme value theory, Embrechts, Klüppelberg and Mikosch (1997).

1.6.1   Coherent risk measures

In order to define coherent risk measure the notion of a risk measure has to be fixed.

Definition

Let Ω denote the set of all states of the world. Assume there are only finitely many states of the world. A risk is a real valued function on Ω and $\mathcal{G}$ is the set of all risks. A risk measure is a real valued function on $\mathcal{G}$.

A risk measure ρ on Ω is called coherent iff

$$\text{For all } X \in \mathcal{G}, \rho(X) \leq \|X^+\|_\infty \tag{1.16}$$

$$\text{For all } X_1 \text{ and } X_2 \in \mathcal{G}, \rho(X_1 + X_2) \leq \rho(X_1) + \rho(X_2) \tag{1.17}$$

$$\text{For all } \lambda \geq 0 \text{ and } X \in \mathcal{G}, \rho(\lambda X) = \lambda \rho(X) \tag{1.18}$$

$$\text{For every subset } A \subset \Omega, X \in \mathcal{G}, \rho(1_A X) \leq \rho(X) \tag{1.19}$$

$$\text{If } X \in \mathcal{G} \text{ is positive and if } \alpha \geq 0 \text{ then } \rho(\alpha + X) = \rho(X) + \alpha \tag{1.20}$$

In Artzner et al. (1997b) it shown that the notion of coherent risk measure is equivalent to the notion of generalized scenarios.

$$\rho_P(X) \;\; = \;\; \sup\{E_P[X^+] | P \in \mathcal{P}\}, \tag{1.21}$$

where $\mathcal{P}$ is a set of probability measures on Ω.

The space we are working with is

$$\Omega \;\; = \;\; \{0, ..., N\}^m.$$

Here N is the largest possible values, as multiples of the basic currency. If $\omega = (\omega(1), .., \omega(m))$, then $\omega(i)$ is the interpreted as the loss in the i-th transaction, if ω is the "state of the world" which is identified with "state of the portfolio".

1.6.2   Capital Definition

As a scenario we choose our observed distribution of $(L_1, .., L_m)$ conditioned on the event that $L = \sum_{k=1}^m L_k > K$ for some large constant K. This constant indicates how the senior management understands under "large losses" for the portfolio. Then a coherent risk measure is defined by

$$\rho(X)_{K,L} \;\; = \;\; E[X|L > K].$$

This is coherent by definition since the measure $P[\cdot|L > K]$ is a probability measure on Ω. Of course this measure is portfolio inherent. The risk factors outside the portfolio, like the asset values are not part of the underlying probability space.

However a straight forward capital definition is then

$$EC_K(III) \;\; = \;\; E[L|L > K].$$

If $K = q_\alpha(L)$, i.e. K is a quantile of the loss distribution, then the map

$$\rho(X) = E[X|q_\alpha(X)]$$

is shown to be a coherent risk measure on any finite Ω, cf. Artzner et al. (1997b).

1.6.3   Contribution to Shortfall-Risk

One main advantage of $EC_K(III)$ is the simple allocation of the capital to the single transaction. The contribution to shortfall risk, CSR is defined by

$$CSR_k \; = \; E[L_k|L > K].$$

That is the capital for a single deal is its average loss in bad situations. Again this is a coherent risk measure on Ω. It is obvious that $CSR_k \leq l_k$. Hence a capital quota of over 100% is impossible, in contrast to the approach based on risk contributions.

1.7   Comparision of the capital allocation methods

We did an analysis on a portfolio of 40 counterparty and based the capital on the 99%-quantile. In table 3 in the appendix the default probabilities and the exposure are reported.

The asset correlation matrix is reported in table 4 in the appendix.

1.7.1   Analytic Risk Contribution

The risk contribution method yield the following contributory economic capital. The first line contains the transaction ID, the second line the analytic derived contributory capital and the third line the same derived in the Monte-Carlo-Simulation. As you see the last two lines are quite close.

Facility ID	1A	2A	3A	4A	5A	6A	7A	8A	9A	10A
Analytic RC	9.92	8.52	8.60	17.86	4.23	2.90	6.19	0.29	2.67	3.45
Monte-Carlo RC	9.96	8.41	8.64	17.93	4.46	2.78	6.06	0.25	2.52	3.39

Facility ID	11A	12A	13A	14A	15A	16A	17A	18A	19A	20A
Analytic RC	1.51	1.87	6.32	18.23	1.51	1.15	2.28	1.51	1.24	0.49
Monte-Carlo RC	1.35	1.85	6.25	18.52	1.45	1.23	2.28	1.60	1.17	0.48

Facility ID	21A	22A	23A	24A	25A	26A	27A	28A	29A	30A
Analytic RC	2.77	0.69	1.43	0.39	3.71	1.90	1.61	4.42	0.58	2.45
Monte-Carlo RC	2.71	0.69	1.44	0.48	3.62	1.86	1.78	4.53	0.60	2.45

Facility ID	31A	32A	33A	34A	35A	36A	37A	38A	39A	40A
Analytic RC	4.27	12.39	0.44	0.98	3.86	5.73	0.80	6.19	3.88	1.08
Monte-Carlo RC	4.29	12.41	0.42	0.88	3.88	5.66	0.79	6.29	3.99	1.06

1.7.2   Simulation procedure

Firstly, the scenarios of the "Ability to pay" at year $1, A^l, l = 1, .., N = $ number of simulations, are generated for all counterparties in the portfolio. For the different types of contributory capital we proceed as follows

Marginal Capital In each realization A^l we consider all losses $L^k(l) :=$ $L - L_k$ in the portfolio without counterparty k for all $k = 1, ..m$. At the end, after N simulations of the asset values we calculate the empirical quantiles $q_\alpha(L^k)$ of each vector $(L^k(1), .., L^k(N))$. The contributory economic capital is then proportional to $q_\alpha(L) - E[L] - q_\alpha(L^k) + E[L^k]$

The performance of this was not satisfactory in a run with even 10.000.000 simulations the single CECs differed quite a lot. Since we are working on an improvement of the simulation procedure we postpone the detailed analysis of this type of contributory economic capital to a forthcoming paper.

Contribution to Shortfall Risk First, the threshold K was set to 150.000.000, which was close to the EC(99%) of the portfolio. In a simulation step where the threshold was exceeded we stored the loss of a counterparty if his loss was positive. After all simulations the average is then easily obtained.

Here we got very stable results for 1.000.000 simulations which can be seen in the following table. Taking 10.000.000 simulations didn't improve the stability significantly.

ID	1A	2A	3A	4A	5A	6A	7A	8A	9A	10A
Run 1	13.22	11.38	10.99	15.87	4.06	2.83	6.00	0.12	2.27	3.10
Run 2	13.64	11.23	11.18	15.75	4.01	2.81	6.89	0.15	2.40	2.95

ID	11A	12A	13A	14A	15A	16A	17A	18A	19A	20A
Run 1	1.41	1.63	5.92	15.49	1.40	1.07	1.95	1.44	0.98	0.48
Run 2	1.33	1.60	6.00	15.34	1.56	1.07	2.10	1.34	1.02	0.47

ID	21A	22A	23A	24A	25A	26A	27A	28A	29A	30A
Run 1	2.29	0.58	1.38	0.33	3.44	1.74	1.36	3.92	0.55	2.36
Run 2	2.35	0.52	1.27	0.35	3.36	1.69	1.25	4.05	0.48	2.23

ID	31A	32A	33A	34A	35A	36A	37A	38A	39A	40A
Run 1	4.09	11.90	0.40	0.85	3.16	5.48	0.79	5.74	3.63	0.93
Run 2	3.98	11.83	0.38	0.82	3.31	5.51	0.76	5.79	3.56	1.04

1.7.3 Comparison

In the present portfolio example the difference between the contributory capital of two different types, namely analytic risk contributions and contribution to shortfall, should be noticed, since even the order of the assets according to their risk contributions changed. The asset with the largest shortfall contributions, 4A, is the one with the second largest risk contribution and the largest risk contributions 14A goes with the second largest shortfall contribution. A view at the portfolio shows that the shortfall distributions is more driven by the relative asset size. Asset 14A has the largest default probability and higher R^2, i.e. systematic risk, than 4A whereas 4A has the second largest exposure and the second largest default probability. Similar observation can be done for the pair building third and fourth largest contributions, asset 1A and 32A. The fifth and sixth largest contribution shows that shortfall risk assigns more capital to the one with larger R^2 since the other two parameter are the same. However this might be caused by statistical fluctuations.

Also the shortfall contribution based on a threshold close to the 99.98% quantile produces the same two largest consumer of capital, namely 4A and 14A.

However, it is always important to bear in mind that these results are still specific to the given portfolio. Extended analysis will be carried out for different types of portfolios in a future study. In these future studies different features might arise. On the lower tranch of the contributory economic capital the two rankings coincide. The lowest is 8A, the counterparty with the lowest correlation (around 13%) to all other members of the portfolio and the smallest default probability, namely 0.0002. The following four lowest capital user also have a default probability of 0.0002 but higher correlation, around 30% to 40%. Counterparty 22A with the sixth lowest capital has a default probability of 0.0006 but a very small exposure and correlations around 20%. Hence both capital allocation methods produce reasonable results.

1.7.4   Portfolio size

The main disadvantage of the simulation based methods are the sizes of the portfolio. For example to get any reasonable number out of the contribution to shortfall risk it is necessary that we observe enough losses in bad cases. Since there are around 1% bad cases of all runs we are left with 10.000 bad scenarios if we had 1.000.000 simulations. Since we have to ensure that each counterparty suffered losses in some of these 10.000 cases we arrive at a combinatorial problem. A way out of this for large portfolios might be to look for capital allocation only to subportfolio instead of an allocation to single counterparties. Since there will be a loss for a subportfolio in most of the bad scenarios, i.e. because of the fluctuation of losses in a subportfolio, the results stabilize with a smaller number of simulations. A detailed analysis of these subportfolio capital allocation for large portfolio will be carried out in a forthcoming paper.

1.8 Summary

We presented three methods to allocate risk capital in a portfolio of loans. The first method is based on the Variance/Covariance analysis of the portfolio. From a mathematical point of view it assumes that the quantile of the loss distribution is a multiple of the variance. This risk contributions are reasonable if the returns are normal distributed. However this is not the case of returns from loans. Since one either obtained the nominal amount of the loan at maturity or one obtains nothing[1] . This binary features motivates the search for other risk measures. One proposed risk measure are the marginal risk contributions, which in our simulation study didn't provide stable results. A third method which also shares some properties of a coherent risk measure in the sense of Artzner et al. (1997b) turned out to be stable for a reasonable number of simulations for a portfolio of 40 loans. The observed differences with the risk contributions were at a first view not very significant. But since even the order of the assets according to their capital usage were changed we look further into some special assets. It turned out that the shortfall contribtuions allocates higher capital to those counterparties with higher exposures. It therefore puts more emphasis to name concentration. However this might be caused by the small size of the portfolio. Shortfall contributions in connection with the definition of shortfall risk prevents of course one phenomena observed for the risk contributions, namely that the capital quota might exceed 100%. The disadvantage of the shortfall contributions is that the computation requires Monte-Carlo-Simulation. This method can be used for allocation of capital to subportfolios or if one is really interested in capital allocation to each single transaction the procedure is restricted to small portfolios.

---

[1]In credit risk management one assumes usually a recovery rate, i.e. a percentage of the exposure that one recovers from defaulted loans. In the present paper this is set to 0.

Appendix

Portfolio

Id	Default Probability	Exposure
1A	0.0024	200,000,000
2A	0.002	200,000,000
3A	0.002	200,000,000
4A	0.0063	146,250,000
5A	0.0013	140,000,000
6A	0.0008	110,000,000
7A	0.002	110,000,000
8A	0.0002	100,000,000
9A	0.0009	100,000,000
10A	0.0013	100,000,000
11A	0.0005	100,000,000
12A	0.0008	100,000,000
13A	0.0024	83,250,000
14A	0.0095	82,500,000
15A	0.0006	81,562,500
16A	0.0004	70,000,000
17A	0.0009	120,000,000
18A	0.0006	62,500,000
19A	0.0006	60,000,000
20A	0.0002	60,000,000
21A	0.0016	55,882,353
22A	0.0006	37,500,000
23A	0.0004	55,000,000
24A	0.0002	55,000,000
25A	0.0017	55,000,000
26A	0.0005	50,000,000
27A	0.001	50,000,000
28A	0.0019	50,000,000
29A	0.0002	50,000,000
30A	0.0012	45,454,545
31A	0.0014	115,000,000
32A	0.0079	44,288,136
33A	0.0002	43,750,000
34A	0.0007	42,000,000
35A	0.0034	37,500,000
36A	0.0031	37,000,000
37A	0.0004	35,000,000
38A	0.0034	35,000,000
39A	0.0031	30,000,600
40A	0.0004	30,000,000

Bibliography

Artzner, P., Dealban, F., Eber, J. and Heath, D. (1997a). Credit risk - a risk special supplement, *RISK MAGAZINE* **7**.

Artzner, P., Dealban, F., Eber, J. and Heath, D. (1997b). Thinking coherently, *RISK MAGAZINE* .

Baestaens, D. and van den Bergh, W. M. (1997). A portfolio approach to default risk, *Neural Network World* **7**: 529–541.

Credit Metrics (1997). *Technical report*, J.P. Morgan & Co.

Embrechts, P., Klüppelberg, C. and Mikosch, T. (1997). *Modelling Extremal Events*, Springer, Berlin.

Fingers, C. (1999). Conditional approaches for creditmetrics portfolio distribution, *CreditMetrics Monitor* .

Kealhofer, S. (1995). Managing default risk in portfolio of derivatives, *Derivative Credit Risk: Advances in Measurement and Management, Renaissance Risk Publications* .

Kealhofer, S. and Crosbie, P. (1997). Modeling portfolio risk, internal document, *Technical report*, KMV Corporation, San Francisco.

Markowitz, H. M. (1952). Portfolio selection, *Journal of Finance* **7**.

Merton, R. (1974). On the pricing of corporate debt: The risk structure of interest rates, *The Journal of Finance* **29**: 449–470.

Overbeck, L. and Stahl, G. (1997). Stochastische Methoden im Risikomanagement des Kreditportfolios, Oehler.

Risk, C. (1997). A credit risk management framework, *Technical report*, Credit Suisse Financial Products.

Schmid, B. (1997). Creditmetrics, *Solutions* **1**(3-4): 35–53.

Sharpe, W. (1964). Capital asset prices: A theory of market equilibrium under conditions of risk, *Journal of Finance* **19**.

Vasicek, O. A. (1997). Credit valuation, *Net Exposure* **1**.

Wilson, T. (1997). Portfolio credit risk (i+ii), *Risk Magazine* **10**.

Table 1.1: Asset correlation

	2A	3A	4A	5A	6A	7A	8A	9A	10A	11A	12A	13A	14A	15A	16A	17A	18A	19A	20A	21A
1A	0.52	0.52	0.47	0.45	0.48	0.48	0.14	0.45	0.44	0.44	0.37	0.50	0.48	0.41	0.44	0.37	0.43	0.39	0.40	0.38
2A		0.52	0.49	0.46	0.49	0.49	0.15	0.45	0.44	0.44	0.38	0.50	0.50	0.42	0.44	0.38	0.45	0.40	0.40	0.41
3A			0.48	0.46	0.48	0.48	0.15	0.46	0.44	0.45	0.38	0.50	0.49	0.42	0.45	0.37	0.43	0.40	0.40	0.37
4A				0.42	0.45	0.46	0.12	0.41	0.41	0.40	0.34	0.46	0.45	0.38	0.41	0.35	0.42	0.36	0.37	0.39
5A					0.42	0.42	0.12	0.40	0.40	0.39	0.33	0.45	0.42	0.37	0.39	0.32	0.38	0.34	0.36	0.31
6A						0.42	0.11	0.42	0.41	0.40	0.34	0.46	0.45	0.38	0.41	0.35	0.42	0.36	0.37	0.40
7A							0.12	0.42	0.41	0.40	0.34	0.46	0.45	0.38	0.41	0.35	0.43	0.36	0.38	0.40
8A								0.13	0.12	0.15	0.16	0.16	0.15	0.17	0.12	0.10	0.11	0.26	0.10	0.04
9A									0.38	0.39	0.33	0.44	0.42	0.37	0.39	0.32	0.38	0.34	0.35	0.32
10A										0.37	0.32	0.43	0.41	0.35	0.37	0.32	0.37	0.33	0.35	0.32
11A											0.33	0.43	0.41	0.37	0.38	0.31	0.36	0.34	0.33	0.29
12A												0.37	0.35	0.31	0.32	0.26	0.31	0.31	0.29	0.25
-13A													0.47	0.40	0.43	0.36	0.42	0.39	0.39	0.37
14A														0.39	0.41	0.36	0.41	0.38	0.36	0.30
15A															0.36	0.30	0.35	0.35	0.32	0.33
16A																0.32	0.37	0.33	0.34	0.31
17A																	0.32	0.33	0.28	0.40
18A																		0.34	0.34	0.28
19A																			0.30	0.34
20A																				0.29

Table 1.2: Asset correlation

	22A	23A	24A	25A	26A	27A	28A	29A	30A	31A	32A	33A	34A	35A	36A	37A	38A	39A	40A
1A	0.28	0.51	0.35	0.46	0.53	0.36	0.49	0.43	0.44	0.45	0.49	0.39	0.32	0.33	0.48	0.40	0.49	0.39	0.47
2A	0.30	0.52	0.35	0.47	0.55	0.37	0.50	0.45	0.45	0.46	0.50	0.40	0.35	0.34	0.49	0.41	0.49	0.39	0.48
3A	0.27	0.52	0.36	0.46	0.54	0.37	0.50	0.44	0.45	0.45	0.50	0.40	0.32	0.33	0.48	0.41	0.50	0.41	0.47
4A	0.29	0.47	0.32	0.43	0.51	0.34	0.45	0.41	0.40	0.42	0.45	0.36	0.33	0.32	0.45	0.37	0.44	0.35	0.45
5A	0.22	0.47	0.34	0.40	0.47	0.33	0.47	0.39	0.40	0.40	0.44	0.35	0.26	0.29	0.43	0.35	0.43	0.38	0.41
6A	0.29	0.47	0.32	0.43	0.51	0.34	0.45	0.41	0.40	0.43	0.45	0.36	0.33	0.32	0.45	0.37	0.44	0.35	0.45
7A	0.29	0.47	0.32	0.44	0.51	0.35	0.46	0.42	0.41	0.43	0.46	0.36	0.33	0.32	0.45	0.37	0.45	0.35	0.45
8A	0.03	0.15	0.11	0.13	0.15	0.08	0.14	0.10	0.13	0.12	0.15	0.13	0.06	0.08	0.14	0.23	0.16	0.16	0.13
9A	0.24	0.45	0.31	0.40	0.47	0.32	0.44	0.38	0.39	0.39	0.43	0.35	0.27	0.29	0.42	0.35	0.43	0.35	0.41
10A	0.24	0.44	0.30	0.39	0.46	0.32	0.43	0.37	0.39	0.38	0.42	0.34	0.27	0.28	0.42	0.35	0.42	0.34	0.40
11A	0.21	0.44	0.30	0.39	0.45	0.31	0.43	0.37	0.38	0.37	0.42	0.35	0.25	0.28	0.41	0.35	0.44	0.35	0.39
12A	0.18	0.38	0.26	0.33	0.39	0.26	0.36	0.31	0.33	0.32	0.36	0.30	0.22	0.24	0.35	0.32	0.37	0.30	0.34
13A	0.26	0.50	0.35	0.44	0.52	0.35	0.48	0.42	0.44	0.43	0.48	0.39	0.30	0.32	0.47	0.40	0.48	0.39	0.45
14A	0.27	0.48	0.33	0.44	0.51	0.34	0.46	0.42	0.42	0.43	0.46	0.37	0.31	0.31	0.45	0.39	0.46	0.37	0.44
15A	0.22	0.42	0.28	0.37	0.43	0.29	0.40	0.35	0.36	0.36	0.40	0.33	0.26	0.27	0.39	0.36	0.41	0.33	0.38
16A	0.24	0.44	0.30	0.39	0.46	0.31	0.42	0.37	0.38	0.38	0.42	0.34	0.28	0.29	0.41	0.34	0.42	0.34	0.40
17A	0.23	0.36	0.25	0.34	0.39	0.27	0.35	0.32	0.32	0.33	0.35	0.28	0.26	0.24	0.35	0.29	0.34	0.27	0.34
18A	0.29	0.42	0.28	0.40	0.47	0.31	0.40	0.38	0.36	0.39	0.41	0.32	0.33	0.30	0.41	0.35	0.40	0.31	0.42
19A	0.20	0.39	0.27	0.35	0.41	0.27	0.38	0.32	0.34	0.34	0.38	0.31	0.25	0.25	0.37	0.39	0.38	0.32	0.36
20A	0.21	0.41	0.27	0.35	0.42	0.30	0.39	0.34	0.36	0.35	0.39	0.30	0.24	0.26	0.38	0.31	0.37	0.30	0.37
21A	0.35	0.35	0.21	0.37	0.44	0.28	0.31	0.36	0.29	0.36	0.35	0.26	0.39	0.28	0.37	0.29	0.33	0.21	0.41
22A		0.26	0.16	0.27	0.32	0.21	0.23	0.26	0.21	0.27	0.25	0.19	0.29	0.20	0.27	0.21	0.24	0.15	0.30
23A			0.36	0.46	0.53	0.37	0.51	0.43	0.46	0.44	0.50	0.40	0.30	0.33	0.48	0.40	0.50	0.41	0.46
24A				0.36	0.49	0.25	0.37	0.29	0.31	0.30	0.34	0.28	0.18	0.22	0.32	0.28	0.34	0.32	0.31
25A					0.31	0.33	0.43	0.40	0.39	0.41	0.44	0.35	0.31	0.30	0.43	0.36	0.43	0.34	0.43
26A						0.39	0.51	0.47	0.46	0.47	0.51	0.40	0.37	0.36	0.51	0.42	0.50	0.40	0.50
27A							0.35	0.32	0.32	0.32	0.35	0.28	0.24	0.24	0.35	0.28	0.34	0.27	0.34
28A								0.41	0.43	0.43	0.48	0.39	0.27	0.32	0.46	0.39	0.48	0.43	0.44
29A									0.37	0.39	0.42	0.33	0.30	0.29	0.41	0.34	0.41	0.32	0.41
30A										0.38	0.43	0.35	0.25	0.28	0.42	0.35	0.43	0.35	0.39
31A											0.42	0.34	0.31	0.30	0.42	0.35	0.42	0.33	0.42
32A												0.38	0.29	0.32	0.46	0.39	0.47	0.38	0.44
33A													0.22	0.25	0.37	0.31	0.39	0.32	0.35
34A														0.23	0.31	0.26	0.28	0.18	0.34
35A															0.31	0.26	0.31	0.24	0.31
36A																0.38	0.46	0.36	0.44
37A																	0.39	0.33	0.31
38A																		0.39	0.37
39A																			0.34

2 Estimating Volatility for Long Holding Periods

Rüdiger Kiesel, William Perraudin and Alex Taylor

2.1 Introduction

The problem of estimating volatility is one of the most important topics in modern finance. Accurate specification of volatility is a prerequisite for modelling financial time series, such as interest rates or stocks, and crucially affects the pricing of contingent claims. Modelling volatility has therefore be widely discussed in the financial literature, see Campbell, Lo and MacKinlay (1997), chapter 12, Shiryaev (1999), chapter 4, or Taylor (1986), chapter 3 for overviews on the subject. The main focus in these studies has been to estimate volatility over short time periods and deduce results for longer period volatility from underlying models.

In this note, we address the problem of estimating volatility over longer time intervals directly. Recently several attempts have been made to examine this problem, most notably work by Andersen (1998), Andersen, Bollerslev, Diebold and Labys (1999), who use intraday observations to estimate the distribution of daily volatility, and Drost and Nijman (1993), Drost and Werker (1996), who consider temporal aggregation of GARCH processes. In contrast to these approaches we do not assume any underlying parametric model for the data generating processes. Our only assumption is that the data generating process is first-difference stationary. The model free approach leads to an estimator, which is insensitive to short-period contamination and only reacts to effects relevant to the time period in question. Applications of the proposed estimator can be found in Cochrane (1988), who used the estimator to obtain a measure of the persistence of fluctuations in GNP, and Kiesel, Perraudin and Taylor (1999), who estimated the long term variability of credit spreads.

Related to our estimation problem are so-called moment ratio tests, which are frequently used to investigate the (weak) efficiency of financial markets, see Campbell et al. (1997), chapter 1, or Pagan (1996) for surveys and Lo and MacKinlay (1988) and Groenendijk, Lucas and de Vries (1998) for applications related to this investigation.

The motivation behind the estimator is as follows. From the assumption that the data generating process x_t is first-difference stationary (i.e. contains a unit root), we obtain from Wold's decomposition (see e.g. Fuller (1996), §2.10) an infinite moving average representation

$$\Delta x_t = x_t - x_{t-1} = \mu + \sum_{j=0}^{\infty} a_j \epsilon_{t-j}. \tag{2.1}$$

Using this representation a result by Beveridge and Nelson (1981) implies that x_t can be represented as the sum of a stationary and a random walk component, i.e

$$x_t = y_t + z_t \tag{2.2}$$

where

$$-y_t = \left(\sum_{j=1}^{\infty} a_j\right)\epsilon_t + \left(\sum_{j=2}^{\infty} a_j\right)\epsilon_{t-1} + \left(\sum_{j=3}^{\infty} a_j\right)\epsilon_{t-2} + \dots \tag{2.3}$$

$$z_t = \mu + z_{t-1} + \left(\sum_{j=0}^{\infty} a_j\right)\epsilon_t, \tag{2.4}$$

with (ϵ_t) a sequence of uncorrelated $(0, \sigma^2)$ random variables.

The long-period behaviour of the variance of the process x_t may differ substantially for processes with representation (2.2). This becomes of particular importance for valuation of contingent claims and, in case of interest rate models, for bond pricing, since the pricing formulae crucially depend on the volatility. Since, in general, the long-term behaviour of the variance of x_t is dominated by the variance of the random walk component, the use of a volatility estimator based on daily time intervals to contingent claims/bonds longer time to maturity may lead to substantial pricing errors. In the next section, we introduce the estimator and discuss some of its properties. We perform Monte Carlo experiments to illustrate the properties of the estimator in section 3. In section 4 we apply it to estimate long holding period variances for several interest rate series. By analysing the quotient of long-term to short-term variances (variance ratio) we can infer the magnitude of the random walk component in the short term interest rate process. This has implications for the appropriate modelling of the short rate and relates to recent results on the empirical verification of various short-term interest rate models, see Bliss and Smith (1998), Chan, Karolyi, Longstaff and Saunders (1992). Section 5 concludes.

2.2   Construction and Properties of the Estimator

We start with a general representation of a first-difference stationary linear process as the sum of a stationary and a random walk component, i.e

$$x_t = y_t + z_t \tag{2.5}$$

with

$$
\begin{aligned}
y_t &= B(L)\delta_t &\tag{2.6}\\
z_t &= \mu + z_{t-1} + \epsilon_t, &\tag{2.7}
\end{aligned}
$$

with $B(L)$ a polynomial in the lag operator $L\delta_t = \delta_{t-1}$, (ϵ_t) uncorrelated, $(0, \sigma^2)$ random variables, and $\mathbb{E}(\epsilon_t \delta_t)$ arbitrary. Such a decomposition implies that $\mathbb{E}_t(x_{t+k}) \approx z_t + k\mu$. In that sense we call z_t the permanent and y_t the temporary component of x_t (compare also Campbell et al. (1997) for a related model and interpretation). This suggests that the long term variability of x_t is also dominated by the innovation variance $\sigma^2_{\Delta z}$ of the random walk component. Utilizing the Beveridge and Nelson (1981) decomposition of a process x_t given by (2.5) one can show that the innovation variance $\sigma^2_{\Delta z}$ is invariant to the particular decomposition of type (2.5) chosen (in particular, only the Beveridge-Nelson decomposition is guaranteed to exist, see also Cochrane (1988)). To make the above arguments on the importance of the innovation variance more precise, consider the $k-$period variability. A standard argument (compare §2.1) shows

$$\boldsymbol{V}ar_t(x_{t+k} - x_t) = k\gamma_0 + 2\sum_{j=1}^{k-1}(k-j)\gamma_j, \tag{2.8}$$

with γ_j the autocovariances of the stationary process $(\Delta x_t) = (x_t - x_{t-1})$. Then

$$\lim_{k\to\infty} \frac{1}{k}\boldsymbol{V}ar_t(x_{t+k} - x_t) = \lim_{k\to\infty}\left(1 + 2\sum_{j=1}^{k-1}\frac{(k-j)}{k}\rho_j\right)\sigma^2_{\Delta x} = S_{\Delta x}(e^{-i0}), \tag{2.9}$$

where ρ_j are the autocorrelations and $S_{\Delta x}(e^{-i\omega})$ is the spectral density function at frequency ω of (Δx_t). A further application of the Beveridge-Nelson decomposition implies

$$S_{\Delta x}(e^{-i0}) = \sigma^2_{\Delta z}. \tag{2.10}$$

Therefore, in order to estimate $\sigma_{\Delta z}^2$ we could use an estimator of the spectral density at frequency zero. However, estimating the spectral density function at low frequencies is extremely difficult and involves a trade-off between bias and efficiency of the estimator (see e.g. Fuller (1996) §7.3 for such estimators and their properties). So, rather than relying on estimators for the spectral density function, we proceed directly with an estimator suggested by (2.8)-(2.10). In particular, (2.8) suggests to replace the autocovariance functions with their sample estimators and then employ well-known limit theorems for the sample autocovariances.

2.2.1   Large Sample Properties

In order to use (2.8), we recall that, under our assumptions, Δx is a covariance stationary process and, as such, has a moving average representation (2.1). Limit theorems for the sample autocovariances of such processes have been studied extensively (see Davis and Resnick (1986), Embrechts, Klüppelberg and Mikosch (1997) §7, Fuller (1996), §6) and we intend to utilize some of these results (much the same way as Lo and MacKinlay (1988) did). Let us start by expressing the basic estimator

$$\bar{\sigma}_k^2 = \frac{1}{Tk} \sum_{j=k}^{T} \left[(x_j - x_{j-k}) - \frac{k}{T}(x_T - x_0) \right]^2 \tag{2.11}$$

in a different form. Define $\hat{\epsilon}_j = x_j - x_{j-1} - \frac{1}{T}(x_T - x_0)$ then (2.11) becomes

$$
\begin{aligned}
\bar{\sigma}_k^2 &= \frac{1}{Tk} \sum_{j=k}^{T} \left[\sum_{l=1}^{k}(x_{j-k+l} - x_{j-k+l-1} - \frac{1}{T}(x_T - x_0)) \right]^2 \\
&= \frac{1}{Tk} \sum_{j=k}^{T} \left[\sum_{l=1}^{k} \hat{\epsilon}_{j-k+l} \right]^2 \\
&= \frac{1}{Tk} \sum_{j=0}^{T-k} \left[\sum_{l=1}^{k} \hat{\epsilon}_{j+l}^2 + 2 \sum_{l=1}^{k-1} \hat{\epsilon}_{j+l}\hat{\epsilon}_{j+l+1} + \ldots + 2\hat{\epsilon}_{j+1}\hat{\epsilon}_{j+k} \right] \\
&= \hat{\gamma}(0) + 2\frac{(k-1)}{k}\hat{\gamma}(1) + \ldots + \frac{2}{k}\hat{\gamma}(k-1) + o\,(.)
\end{aligned}
$$

where

$$\hat{\gamma}(h) = \frac{1}{T} \sum_{j=0}^{T-h} \hat{\epsilon}_j \hat{\epsilon}_{j+h}$$

and $o\,(.)$ specifies an error term in probability depending on the distribution of the innovations. Define the vector $\hat{\boldsymbol{\gamma}} = (\hat{\gamma}(0), \ldots, \hat{\gamma}(k-1))'$, then we can write

$$\bar{\sigma}_k^2 = l'\hat{\gamma} + o(.) \tag{2.12}$$

with l the k-dimensional vector $l = (1, 2\frac{(k-1)}{k}, \ldots, \frac{2}{k})'$. We therefore can use limit theorems on the asymptotic distribution of $\hat{\gamma}$ to deduce the asymptotic distribution of our estimator $\bar{\sigma}_k^2$. These limit theorems depend crucially on the distribution of the innovations ϵ in (2.1). If $I\!\!E(\epsilon^4) < \infty$, the limit distribution of $\hat{\gamma}$ is Gaussian, see e.g. Brockwell and Davis (1991), §7.3,§13.3, Fuller (1996) §6.3. If $I\!\!E(\epsilon^4) = \infty, \sigma_\epsilon^2 < \infty$ (and further regularity conditions are satisfied), the limit distribution consists of a stable random variable multiplied by a constant vector, see Davis and Resnick (1986) and Embrechts et al. (1997) §7.3 for details. Hence, in the first case the asymptotic distribution of $\bar{\sigma}_k^2$ will be Gaussian, while in the second case it will asymptotically be distributed according to a stable law.

2.2.2   Small Sample Adjustments

In small samples, the estimator (2.11) exhibits a considerable bias. To discuss possible adjustments we assume that the data generating process is a pure unit root process, i.e. equation (2.5) becomes

$$\Delta x_t = \Delta z_t = \mu + \epsilon_t \tag{2.13}$$

with (ϵ_t) uncorrelated $(0, \sigma^2)$ random variables. So we can write the numerator of the estimator (2.11)

$$\begin{aligned}
N_\sigma &= \sum_{j=k}^{T} \left((x_j - x_{j-k}) - \frac{k}{T}(x_T - x_0) \right)^2 \\
&= \sum_{j=k}^{T} \left(k\mu + \sum_{\nu=0}^{k-1} \epsilon_{j-\nu} - \frac{k}{T}\left(T\mu + \sum_{\nu=0}^{T-1} \epsilon_{T-\nu} \right) \right)^2 \\
&= \sum_{j=k}^{T} \left(\sum_{\nu=n-k+1}^{n} \epsilon_\nu - \frac{k}{T}\sum_{\nu=1}^{T} \epsilon_\nu \right)^2.
\end{aligned}$$

Defining $Z_{j,k} = \sum_{\nu=j-k+1}^{j} \epsilon_\nu$ and using the fact that the ϵ_ν are uncorrelated we get

$$\begin{aligned}
I\!\!E(N_\sigma) &= \sum_{j=k}^{T} \left(I\!\!E(Z_{j,k}^2) - \frac{2k}{T} I\!\!E(Z_{j,k} Z_{T,T}) + \frac{k^2}{T^2} I\!\!E(Z_{T,T}^2) \right) \\
&= I\!\!E(\epsilon^2) \sum_{j=k}^{T} \left(k - \frac{2k^2}{T} + \frac{k^2}{T} \right) = \sigma^2(T - k + 1)(T - k)\frac{k}{T}.
\end{aligned}$$

So in order to get an unbiased estimator for σ^2 using the quantity N_σ we have to multiply it by

$$\frac{T}{k(T-k+1)(T-k)},$$

which is just the adjustment proposed by Cochrane (compare Cochrane (1988)) and leads to

$$\hat{\sigma}_k^2 = \frac{T}{k(T-k)(T-k+1)} \sum_{j=k}^{T} \left[(x_j - x_{j-k}) - \frac{k}{T}(x_T - x_0) \right]^2. \quad (2.14)$$

If we assume that the innovations in (2.13) are uncorrelated and the fourth moment exists, we can use the asymptotic equivalence of the estimators (2.11) and (2.14) to deduce the weak convergence[1]

$$\sqrt{T}(\hat{\sigma}_k^2 - \sigma^2) \Rightarrow N(0, \sigma^4((2k^2+1)/3k)). \quad (2.15)$$

If, however, the last existing moment of the innovations in (2.13) is of order α, where $2 < \alpha < 4$, i.e the variance exists, but the fourth moment is infinite, we have the weak convergence

$$C(T,\alpha)\hat{\sigma}_k^2 \Rightarrow \sqrt{k}S, \quad (2.16)$$

where S is a stable random variable with index $\alpha/2$ and $C(T,\alpha)$ a constant depending on the T and the tail behaviour of the innovations, which is related to the index α. (The relevant asymptotic result for the autocovariances is Theorem 2.2 in Davis and Resnick (1986), where the exact values of the constants to be used to construct the vector l in (2.12) can be found). If we drop the assumption (2.13), the limit laws remain of the same type. However, the variances change considerably since they depend on the autocovariances of the process.[2]

2.3 Monte Carlo Illustrations

In this section, we illustrate our estimating procedure using simulated time series. We consider three basic settings of first-difference stationary sequences with representation (2.1). First, as a benchmark case,

---

[1] We denote weak convergence by "$\Rightarrow$".

[2] For instance in case the $\mathbb{E}(\epsilon^4) = \eta\sigma^4$ we have $\lim_{n \to \infty} \boldsymbol{C}ov(\gamma(\hat{p})\gamma(\hat{q})) = (\eta - 3)\gamma(p)\gamma(q) + \sum_{k=-\infty}^{\infty}[(\gamma(k)\gamma(k-p+q) - \gamma(k+q)\gamma(k-p)]$ and thus $(\hat{\gamma}_1, \dots, \hat{\gamma}_k)' \sim N((\hat{\gamma}_1, \dots, \hat{\gamma}_k)', T^{-1}V)$, where V is the covariance matrix (see Brockwell and Davis (1991), §7.3). This implies an asymptotic standard normal distribution of $\hat{\sigma}_k^2$ with variance $l'Vl$.

	$\hat{\sigma}^2_k$ [a]	s.e.	s.e.	$\hat{\sigma}^2_k$ [b]	s.e.	s.e. [c]	$\hat{\sigma}^2_k$ [d]	s.e.	s.e. [e]
w [f]	0.999	0.048	0.048	0.860	0.039	0.040	1.169	0.057	0.059
m	0.999	0.096	0.094	0.835	0.077	0.078	1.120	0.115	0.117
q	0.997	0.167	0.163	0.830	0.137	0.136	1.206	0.199	0.202
y	0.991	0.347	0.333	0.822	0.288	0.278	1.212	0.422	0.411

Table 2.1: Model with i.i.d. Gaussian innovations

[a]model: $\Delta x_t = \epsilon_t$ with $\epsilon_t \sim N(0,1)$ and $\sigma^2_{\Delta z} = 1$. First s.e. column are always Monte-Carlo, second s.e. column are asymptotic s.e. assuming existence of the fourth moment.

[b]model: $\Delta x_t = a\Delta x_{t-1} + \epsilon_t$ with $\epsilon_t \sim N(0,1)$ and $\sigma^2_{\Delta z} = \left(\frac{1}{1-a}\right)^2 \sigma^2_\epsilon$, here $\sigma^2_{\Delta z} = 0.826$

[c]Adjusted for AR(1)-covariance structure

[d]model: $\Delta x_t = \epsilon_t + a\epsilon_{t-1}$ with $\epsilon_t \sim N(0,1)$ and $\sigma^2_{\Delta z} = (1-a)^2 \sigma^2_\epsilon$, here $\sigma^2_{\Delta z} = 1.21$

[e]Adjusted for MA(1)-covariance structure

[f]w=week, m=month, q=quarter, y=year

we consider a pure random walk with representation as in (2.13). To study the effect of non-zero autocovariances of the series (Δx) on the asymptotic standard error, we simulate two further series, namely a sequence, whose first-difference follows an autoregressive model of order one (AR(1)-model) implying an infinite order moving average representation and on the other hand, a sequence, which has first-differences allowing a moving average representation of order one (MA(1)).

These settings imply that the error terms in (2.5) are perfectly correlated. The AR-model corresponds to a 'small' random walk component (in our setting it accounts for roughly 70% of variability of (x_k) in (2.5)). The MA-model, on the other hand, corresponds to a 'large' random walk component, the innovation variance of the random walk component (z_k) in (2.5) is larger (due to dependence) than the innovation variance of the series (x_k).

For each of these series, we consider three types of innovation process. As a standard model we consider i.i.d. Gaussian innovations. Then we investigate the effect of heavy-tailed innovations using i.i.d. Student $t(3)$ innovations, and finally to discuss (second order) dependence we use GARCH(1,1)-innovations. Each experiment consisted of generating a series of length 3000 (with coefficients in line with coefficients obtained performing the corresponding ARIMA (-GARCH) for the series used in §4) and was repeated 5000 times. We report the mean of long-period volatility estimators for periods of length $k = 5, 20, 60, 250$ (weeks, month,

$\hat{\sigma}^2_k$ [a]	s.e.	s.e.	$\hat{\sigma}^2_k$ [b]	s.e.	s.e.[c]	$\hat{\sigma}^2_k$ [d]	s.e.	s.e.[e]
w^f	2.980	0.962	0.144	2.555	0.644	0.120	3.507	1.388	0.1779
m	2.977	0.983	0.283	2.483	0.667	0.237	3.602	1.458	0.349
q	2.970	1.023	0.490	2.467	0.753	0.490	3.618	1.467	0.605
y	2.992	1.406	1.000	2.464	1.107	0.834	3.621	1.868	1.234

Table 2.2: Model with i.i.d. $t(3)$ innovations

---

[a] model: $\Delta x_t = \epsilon_t$ with $\epsilon_t \sim t(3)$ and $\sigma^2_{\Delta z} = 3$. First s.e. column are always Monte-Carlo, second s.e. column are asymptotic s.e. assuming existence of the fourth moment.

[b] model: $\Delta x_t = a\Delta x_{t-1} + \epsilon_t$ with $\epsilon_t \sim t(3)$ and $\sigma^2_{\Delta z} = \left(\frac{1}{1-a}\right)^2 \sigma^2_\epsilon$, here $\sigma^2_{\Delta z} = 2.479$

[c] Asymptotic standard error, adjusted for AR(1)-covariance structure

[d] model: $\Delta x_t = \epsilon_t + a\epsilon_{t-1}$ with $\epsilon_t \sim t(3)$ and $\sigma^2_{\Delta z} = (1-a)^2 \sigma^2_\epsilon$, here $\sigma^2_{\Delta z} = 3.63$

[e] Asymptotic standard error, adjusted for MA(1)-covariance structure

[f] w=week, m=month, q=quarter, y=year

quarters, years) together with standard errors (s.e.) computed from the Monte-Carlo simulations and according to the asymptotic results for an underlying pure unit root process with an existing fourth moment.

$\hat{\sigma}^2_k$ [a]	s.e.	s.e.	$\hat{\sigma}^2_k$ [b]	s.e.	s.e.[c]	$\hat{\sigma}^2_k$ [d]	s.e.	s.e.[e]
w^f	4.078	0.278	0.192	3.505	0.237	0.161	4.770	0.324	0.237
m	4.066	0.437	0.378	3.390	0.370	0.315	4.887	0.528	0.466
q	4.037	0.710	0.653	3.348	0.595	0.545	4.897	0.871	0.806
y	4.004	1.442	1.333	3.323	1.187	1.113	4.903	1.767	1.645

Table 2.3: Model with GARCH(1,1) innovations

---

[a] model: $\Delta x_t = \epsilon_t$ with $\epsilon_t \sim GARCH(1,1)$ and $\sigma^2_{\Delta z} = 0.004$. First s.e. column are always Monte-Carlo, second s.e. column are asymptotic s.e. assuming existence of the fourth moment.

[b] model: $\Delta x_t = a\Delta x_{t-1} + \epsilon_t$ with $\epsilon_t \sim GARCH(1,1)$ and $\sigma^2_{\Delta z} = \left(\frac{1}{1-a}\right)^2 \sigma^2_\epsilon$, here $\sigma^2_{\Delta z} = 0.003306$

[c] Adjusted for AR(1)-covariance structure

[d] model: $\Delta x_t = \epsilon_t + a\epsilon_{t-1}$ with $\epsilon_t \sim GARCH(1,1)$ and $\sigma^2_{\Delta z} = (1-a)^2 \sigma^2_\epsilon$, here $\sigma^2_{\Delta z} = 0.0484$

[e] Adjusted for MA(1)-covariance structure

[f] w=week, m=month, q=quarter, y=year

In line with the asymptotic consistency of the estimator $\hat{\epsilon}_k{}^2$ (compare

2.8) the estimated value converges towards the true value of the innovation variance of the random walk component in all cases. For Gaussian and GARCH innovation (cases for which the appropriate limit theory holds) the asymptotic standard errors are in line with the observed Monte Carlo errors. As expected the asymptotic standard errors (calculated under the assumption of an existing fourth moment) become unreliable for heavy tailed innovation, i.e. simulations based on $t(3)$ innovations.

Since for shorter series the asymptotic standard error becomes unreliable we also tested various bootstrap based methods. Motivated by the application we have in mind we concentrated on series with length 1000 and standard normal or GARCH innovations. It turned out, that fitting a low-order AR-model to the simulated time series and resampling from the residuals produced satisfactory bootstrap standard errors.

Model	lag 60			lag 250		
	$\hat{\sigma}_k^2$ [a]	B-s.e.	A-s.e.	$\hat{\sigma}_k^2$	B-s.e.	A-s.e.
RW[b]	0.950	0.263	0.286	1.015	0.359	0.583
AR(1)	0.820	0.277	0.253	0.9314	0.668	0.823
MA(1)	1.199	0.349	0.363	1.270	0.816	0.841
RW[c]	3.886	1.163	1.117	3.997	2.634	2.366
AR(1)	3.282	0.960	0.952	3.041	1.887	1.926
MA(1)	4.702	1.311	1.395	4.814	2.823	2.946

Table 2.4: Bootstrap estimates of standard errors

[a]All on a time series of length 1000 with 5000 bootstrap resamples, parameters chosen as above
[b]standard Normal innovations
[c]GARCH-Innovations, values multiplied by 10^3

2.4 Applications

Empirical comparisons of continuous-time models of the short-term interest rate have recently been the focus of several studies, see e.g. Bliss and Smith (1998), Broze, Scaillet and Zakoian (1995), Chan et al. (1992), Dankenbring (1998). In these studies the general class of single-factor diffusion models

$$dr = (\mu - \kappa r)dt + \sigma r^\gamma dW, \qquad (2.17)$$

with constant coefficients and W a standard Brownian motion has been compared. We will consider the subclass, where we restrict the parameter γ to take one of the values $0, 1/2$ or 1, so e.g. the Vasicek and the Cox-Ingersoll-Ross model are included. The discrete-time analog of this model class is

$$r_t - r_{t-1} = \alpha + \beta r_{t-1} + \epsilon_t \qquad (2.18)$$
$$I\!E(\epsilon_t | F_{t-1}) = 0, \quad I\!E(\epsilon_t^2 | F_{t-1}) = \sigma^2 r_{t-1}^{2\gamma},$$

with F_t the information set at time t. A model like this will generate a time series within our framework if $\beta = 0$. If we focus on the unconditional long-term variance a standard calculation shows, that we have the following asymptotic relations (under $\beta = 0$)

$$\gamma = 0 \quad \boldsymbol{V}ar(r_t) \sim t$$
$$\gamma = \tfrac{1}{2} \quad \boldsymbol{V}ar(r_t) \sim t^2$$
$$\gamma = 1 \quad \boldsymbol{V}ar(r_t) \sim e^{ct}$$

(c a constant). Using the Cochrane-type estimator we can compare the observed long-term variances with variances predicted from the model setting. We apply this idea to three short-term (7 day-maturity) interest rate series. The rates we use are US EURO-DOLLAR (with 3512 observations from 01.01.85 – 18.06.98), UK EURO-POUND (with 3401 observations from 01.01.85 – 13.01.98), and German EURO-MARK (with 1222 observations from 09.01.95 – 14.09.99).

rate	$\hat{\sigma}_1^2$	$\hat{\sigma}_5^2$	$\hat{\sigma}_{20}^2$	$\hat{\sigma}_{60}^2$	$\hat{\sigma}_{250}^2$
US EURO-DOLLAR	0.0537	0.0438	0.0149	0.0077	0.0092
	(0.0055[a])	(0.0092)	(0.0107)	(0.0022)	(0.0051)
UK EURO-POUNDS$	0.0439	0.0293	0.0189	0.0169	0.0212
	(0.0051)	(0.0076)	(0.0123)	(0.0080)	(0.0118)
GER EURO-MARK$	0.0059	0.0048	0.0015	0.0013	0.0018
	(0.0031)	(0.0029)	(0.0009)	(0.0008)	(0.0008)

Table 2.5: Short rate volatilities

[a]For lags 1 to 20 s.e are based on asymptotic calculations, for lags 60 and 250 s.e. are bootstrap based

To ensure the validity of the assumption $\beta = 0$ we performed various tests for unit roots and stationarity[3]. For all series we can't reject the

[3]For the unit root tests we used the augmented Dickey-Fuller and Phillips-Perron procedures and for testing stationarity the Kwiatkowski-Phillips-Schmidt-Sin test (see Maddala and Kim (1998) chapters 3 and 4 for a description and discussion of these tests)

presence of a unit root at a 10% significance level, whereas stationarity of the series is rejected at the 1% significance level. Applying these tests again to the first-difference of the series indicated no evidence of a unit root in the differenced series. The combination of these test results allows us to conclude the series should be modelled as first-difference stationary and fit into our framework.

We report the results for the interest series in table (2.5). From a model-free point of view (that is within the general framework (2.5)) these results indicate, that using the one-day volatility estimate will seriously overestimate longer term volatility.

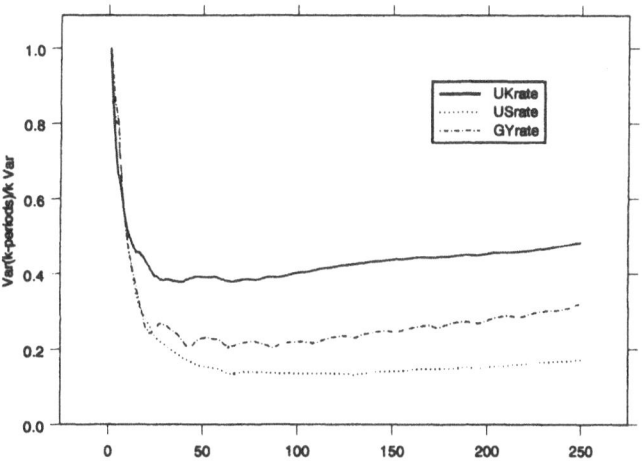

Figure 2.1: Variance-Ratios for short-term interest rates

Turning to the question of modelling short-term interest rates within the class of one-factor diffusion models we calculate and plot the ratio of the volatility calculated over a longer holding period to that calculated over one day multiplied by k (see figure 1). For all rates considered the ratios are downward slopping for short holding periods (the mean-reverting component dies off). After a period of stability the variance ratio begin to increase linearly showing a behaviour roughly in line with the asymptotics of a Cox-Ingersoll-Ross model.

2.5   Conclusion

We presented a non-parametric method to estimate long-term variances and the magnitude of the unit root process in various interest rates. Our results suggest that calculating long-term variances on the basis of short-term variance estimates will overestimate long-term variances. Our results further indicate that within the one-factor diffusion short rate model class square-root type processes model the behaviour of long-term variances of short rates best. Models, which assume that the short rate follows a mean-reverting process and thus omit a unit root component in the data generating process will lead to an underestimating of long-term variances, since for longer time horizons the unit-root component of the interest-rate process becomes dominant. Our findings support a model of Cox-Ingersoll Ross type without a mean-reverting component.

Bibliography

Andersen, T., Bollerslev, T., Diebold, F. and Labys, P. (1999). The distribution of exchange rate volatility, *Technical report*, Wharton School, University of Pennsylvania, Financial Institutions Center.

Andersen, T.G.and Bollerslev, T. (1998). DM-Dollar volatility: Intraday activity patterns, macroeconomic announcements, and longer-run dependencies, *Journal of Finance* **53**: 219–265.

Beveridge, S. and Nelson, C. (1981). A new approach to decomposition of economic time series into permanent and transitory components with particular attention to measurement of the business cycle, *J. Monetary Economics* **7**: 151–174.

Bliss, R. and Smith, D. (1998). The elasticity of interest rate volatility – Chan, Karolyi,Longstaff and Sanders revisited, *Journal of Risk* **1**(1): 21–46.

Brockwell, P. and Davis, R. (1991). *Times series: Theory and methods*, 2 edn, Springer.

Broze, L., Scaillet, O. and Zakoian, J. (1995). Testing for continuous-time models of the short-term interest rate, *J. Empirical Finance* **2**: 199–223.

Campbell, J., Lo, A. and MacKinlay, A. (1997). *The econometrics of financial markets*, Princeton University Press.

Chan, K., Karolyi, G., Longstaff, F. and Saunders, A. (1992). An empirical comparison of alternative models of the short-term interest rates, *Journal of Finance* **47**: 1209–1228.

Cochrane, J. (1988). How big is the random walk in GNP, *J. Political Economics* **96**(51): 893–920.

Dankenbring, H. (1998). Volatility estimates of the short term interest rate with application to german data, *Technical report*, Graduiertenkollog Applied Microeconomics, Humboldt- and Free University Berlin.

Davis, R. and Resnick, S. (1986). Limit theory for the sample covariance and correlation functions of moving averages, *Annals of Statistics* **14**(2): 533–558.

Drost, F. and Nijman, T. (1993). Temporal aggregation of GARCH processes, *Econometrica* **61**: 909–927.

Drost, F. and Werker, B. (1996). Closing the GARCH gap: Continuous time GARCH modeling, *Journal of Econometrics* **74**: 31–57.

Embrechts, P., Klüppelberg, C. and Mikosch, P. (1997). *Modelling extremal events*, Springer.

Fuller, W. (1996). *Introduction to statistical time series*, John Wiley & Sons.

Groenendijk, P., Lucas, A. and de Vries, C. (1998). A hybrid joint moment ratio test for financial time series, *Technical report*, Vrije Universiteit, Amsterdam.

Kiesel, R., Perraudin, W. and Taylor, A. (1999). The structure of credit risk, *Technical report*, Birkbeck College.

Lo, A. and MacKinlay, A. (1988). Stock market prices do not follow random walks: Evidence from a simple specification test, *Review of Financial Studies* **1**(1): 41–66.

Maddala, G. and Kim, I. (1998). *Unit root, cointegration, and structural change Themes in moderen econometrics*, Cambridge University Press.

Pagan, A. (1996). The econometrics of financial markets, *J. Empirical Finance* **3**: 15–102.

Shiryaev, A. (1999). Essentials of stochastic finance, *Advanced Series of Statistical Science & Applied Probability* **3**. World Scientific.

Taylor, S. (1986). *Modelling financial time series*, J. Wiley & Sons.

3 A Simple Approach to Country Risk

Frank Lehrbass

3.1 Introduction

It had been a widely held belief that the debt crisis of the 1980s was over when the Mexican crisis at the end of 1994, the Asian crisis in 1997, and the Russian crisis in 1998 made clear that highly indebted developing countries remain vulnerable. Hence, investment in such countries is still risky and should be assessed properly. A definition of country risk is as follows of the Basel Committee (1999)[p.7]: "Country or sovereign risk encompasses the entire spectrum of risks arising from the economic, political and social environments of a foreign country that may have potential consequences for foreigners debt and equity investments in that country". Note that two markets are mentioned. This paper focusses on the link between these markets. This paper is not concerned with predicting crises. With respect to forecasting there is ample literature available which was recently reviewed by Somerville and Taffler (1995) and Kaminsky, Lizondo and Reinhart (July 1997). Furthermore there is the issue of risk dependencies between different countries that has been emphasized recently of the Basel Committee (1999)[p.7]: "Banks need to understand the globalisation of financial markets and the potential for spillover effects from one country to another". In order to adress this issue as well recent techniques from derivative pricing are applied assuming that market prices of a country's external debt are derived from the prices of other liquid financial instruments. The approach chosen allows the integration of "market and credit risk" and enables portfolio-wide assessment of country risk incorporating correlations.

The theoretical results of the paper are applied to real world data. The focus is on Eurobonds issued by the sovereign of a country. In pricing Eurobonds the market performs a daily up-to-date judgement on the issuer's ability to service his foreign currency liabilities. Of course, other factors such as the default-riskless yield curve of the denomination currency are taken into account as well. Thus, there is a potential source of information concerning country risk. In the sequel the task of extracting this information from market prices is approached. But before possible objections have to be discussed:

One might doubt whether the focus on Eurobonds of sovereign issuers is adequate for judging country risk as defined above. Pragmatically speaking the debt of other foreign borrowers such as foreign corporates is not that actively traded if at all. Hence, data availability dictates this choice. Economically speaking the inability of the sovereign to service foreign currency liabilities can be considered as a sufficient condition for the corresponding inability of the corporates ruled by this sovereign. Of course this condition is not a necessary one, because a coporate may be inable to service foreign debt although the sovereign still is.

Finally, the relevance of the Eurobond market for other debt such as bank loans may be questioned. Facing the grown-up volume of the Eurobond market it is no longer possible to exclude bonds from the general treatment of a country's debt in crisis. Hence, the "comparability of treatment" may diminish the difference between Eurobonds and other forms of debt. Hence, experience from the loan markets will be used for an investigation of the Eurobond-market.

3.2   A Structural No-Arbitrage Approach

In order to extract country risk related information that is condensed into the market prices of Eurobonds a so-called "Structural Model" will be developed in the sequel. This model copes simultaneously with default and interest-rate risk.

3.2.1   Structural versus Reduced-Form Models

Recent academic literature has established the two categories "Structural and Reduced-Form Models". For a discussion of these approaches and literature see Duffie and Lando (1997)[Section 1.2]. Their findings are summarized:

In structural models default occurs when the assets of the issuer have fallen to a sufficiently low level relative to the liabilities, in a sense that varies from model to model. As a rule the existing academic literature models the evolution of the assets of the issuer as a continuous stochastic process and, hence, default occurs not with a "bang but with a whimper". At least two exceptions should be mentioned: Zhou (March 1997) uses a so-called jump-diffusion process for the assets of the issuer, whereas Duffie and Lando (1997) keep the continuous process assumption but presume that investors are uncertain about the current level of the issuer's assets.

The other category of models is called reduced-form. These take as given primitives the fractional recovery of bonds at default, as well as a stochastic intensity for default. Because the default triggering mechanism is not modelled explicitly the only hedge available is a counter position in the same market and not across markets.

Since the structural approach allows for in- and across market hedging and, hence, allows for relative value pricing between markets, it is preferred.

3.2.2   Applying a Structural Model to Sovereign Debt

One might question whether structural models - that have been applied mainly to corporate debt - can be transferred to sovereign debt. Somerville and Taffler (1995)[p.284] point out that the "emergence of arrears ... is in effect the macroeconomic counterpart to corporate failure". Although in the sequel this analogy will turn out to be helpful, it should not be pressed too far. On the one hand it is nearly impossible to conduct reorganizations and to establish covenants with respect to sovereigns, while on the other hand, even existing contractural agreements are very difficult to enforce as will be discussed in more detail below.

It may be doubted whether it is sufficient to rely on financial market data. The explicit inclusion of political factors might appear necessary. De Haan, Siermann and Van Lubek (1997)[p.706] state that "the influence of political factors is discounted in macroeconomic variables". Applying this argument to the relation between daily market prices and monthly macroeconomic figures a sole reliance on financial market data seems defendable. Hence, a free-riding approach to the information processing capabilities of financial markets is applied. Of course there is no presumption towards full informational efficiency as was shown to be impossible by Grossman and Stiglitz (1980). Practicioners agree upon that the evolution of the stock market mirrors the future prospects of the respective country.

In a structural model default occurs when the assets of the issuer have fallen to a sufficiently low level relative to the liabilities. These concepts have to be re-defined in the case of a sovereign issuer. The discounted future stream of the country's GDP is a good candidate for giving meaning to "the assets of the issuer". Since profits are part of GDP (e.g. Branson (1979)[p.21]) part of the discounted future stream of the country's GDP is traded on a daily basis in the stock market. Hence, the equity index of a country is one available proxy for the discounted future stream of the country's GDP. However it should be emphasized that the quality of this proxy is rather dubious because the construction of the available equity indices varies from country to country and the market capitalization does so too. Furthermore the relative weights of the major components of GDP by type of income vary as well. Nevertheless the equity index of a country contains valuable information concerning the future prospects of a country.

Having determined the notion of the assets of the issuer, the relative value of the liabilites may be extracted from Eurobond market prices during a so-called calibration period.

3.2.3   No-Arbitrage vs Equilibrium Term Structure

It has been mentioned that the price of a Eurobond captures not only the inherent risk of default of the issuer, but also the term structure uncertainty of the default-riskless yield curve. The model by Longstaff and Schwartz (1995) copes with both factors simultaneously. Longstaff and Schwartz give pricing formulae for default-risky fixed and floating rate debt. The term structure model used by Longstaff and Schwartz is the equilibrium model by Vasicek (1977). This has the severe effect that the model does not match the current riskless zero yield-curve by construction. Instead the user has to put in the market price of interest rate risk and - as a rule - encounters deviations between the model and the actual riskless zero yield-curve. Therefore it is desirable to replace this equilibrium model by a so-called no-arbitrage model. Due to the work of Hull and White (1990) there is the no-arbitrage version of the equilibrium model: The so-called "Extended Vasicek". This will be used in the sequel.

The time-invariant barrier triggering default, that is used by Longstaff and Schwartz, is replaced by its risklessly discounted level and, hence, no longer fixed. This approach goes back to Schönbucher (November 1997), who proves a closed-form solution for a model thus modified and uses the no-arbitrage model by Ho and Lee (1986). This term-structure model generates upward sloping yield curves only. In contrast the extended Vasicek is capable of generating inverse yield curves as has been shown by Schlögl and Sommer (1994).

3.2.4 Assumptions of the Model

Consider a zero-coupon Eurobond issued by sovereign S denominated in currency of country L with a bullet repayment of the principal and no optional features. There are frictionless markets with continuous trading for the equity-index and currency of the sovereign S and for riskless zero coupon bonds of the country L for any maturity. Hence, the so-called equivalent martingale measure exists as has been shown by Harrison and Kreps (1979). An immediate consequence is that the model does not need any information concerning the risk premia for both term structure and default risk. Hence, for pricing purposes the dynamics are formulated for an artificial risk-neutral world. Note that the resulting formula is valid for any "world" including the real one since it is essentially based on a no-arbitrage reasoning. One more consequence is that specifying the zero bond dynamics of the default-riskless zero coupon bonds it suffices to determine the volatility function as has been shown by Heath, Jarrow and Morton (1992). The volatility at time t of a default-riskless zero coupon bond maturing at time T with price $B(t,T)$ is assumed to be given by the following deterministic function $\nu(t,T)$:

$$\nu(t,T) \stackrel{\text{def}}{=} \frac{\sigma}{\lambda}(1 - \exp{(-\lambda(T-t))}) \tag{3.1}$$

The two parameters σ and λ are identifiable as parameters of the extended Vasicek model (mean reverting process with fitting function 'time dependent mean', σ is the volatility of the so-called short-rate and λ determines the force that pulls the short-rate towards the time-dependent-mean). The SDE for the short rate is:

$$dr = \lambda(mean(t) - r(t))dt - \sigma dw_1 \tag{3.2}$$

The formula for the time dependent mean needs notation for the instantaneous forward rate for time t as it derives from the actual discount curve. This forward rate is denoted by $f(0,t)$:

$$mean(t) \stackrel{\text{def}}{=} \frac{\partial f(0,t)}{\partial t}\frac{1}{\lambda} + f(0,t) + \frac{\sigma^2(1 - \exp{(2\lambda t)})}{2\lambda^2} \tag{3.3}$$

The dynamics of the zero coupon bond's prices are governed by the following stochastic differential equation (=:SDE), where the default-free interest rate for an instantaneous horizon is signified by $r(t)$ and the increments of a Brownian motion by dw_1:

$$\frac{dB(t,T)}{B(t,T)} = r(t)dt + \nu(t,T)dw_1 \tag{3.4}$$

The equity-index of the borrowers country *expressed in units of country L's currency* is denoted by V(t) and assumed to evolve according to the following SDE:

$$\frac{dV(t)}{V(t)} = r(t)dt + \rho\theta dw_1 + \theta\sqrt[2]{1-\rho^2}dw_2 \tag{3.5}$$

The differentials dw_i (i=1,2) are assumed to be independent. Note that the dynamics for the equity-index of the borrower's country expressed in units of country L's currency fit into usual market practice. Currency derivatives are valued assuming a geometric Brownian motion for the exchange rate. The same is done for equity indices. It is known that the product of two geometric Brownian motions is again a geometric Brownian motion. The correlation between the continuous returns of the equity index $V(t)$ and the changes in the default-free short rate $r(t)$ under the risk-neutral measure is given by the constant $-\rho$. The negative sign is due to the fact that interest rates and zero bond prices move in opposite directions.

If $\rho = 0$ the equity-index of the borrowers country expressed in units of country L's currency evolves as in the model of Merton (1973). If in addition the interest rate were constant the dynamics of the equity-index would reduce to that of Black and Scholes (1973) - its volatility given by the constant θ.

The default threshold is given by the following time-variant and stochastic level $\kappa(t,T)$:

$$\kappa(t,T) \stackrel{\text{def}}{=} B(t,T)K \tag{3.6}$$

If this threshold is crossed before maturity by the the equity-index of the borrower's country, one unit of the defaultable bond is replaced by $(1-w)$ units of a default-riskless zero bond with the same maturity date. The parameter w is known as "write down" and $(1 - w)$ as "recovery rate". The write down w is assumed to be deterministic. Note that the default triggering variable "equity-index of the borrowers country expressed in units of country L's currency" incorporates information from the equity and currency markets simultaneously. The fact that there is a connection between creditworthiness and both the currency and stock markets has already been mentioned by Erb, Harvey and Viskanta (1995),

Obviously, this definition of the default event is technical. Other definitions are possible but in general do not allow for the calculation of a closed-form solution for the arbitrage-free price of a Eurobond. The definition of the default as being triggered by the first "touch down" of the equity index is due to the derivatives-pricing technology applied.

The loss occurring in default is similar to the consequences of a debt-bond-swap performed first in 1987 with respect to the sovereign Mexico: Outstanding debt was written off by the lenders; in exchange they received $(1 - w)$ units of newly issued Mexican debt. The principal was collateralized by T-Bonds. Summing up the effects of a debt-bond-swap: The defaulted bond is replaced by $(1 - w)$ units of a default-riskless zero bond with the same maturity date. Of course, it is not claimed that any debt crisis will be handled by such a debt-bond-swap. Instead, this debt-bond-swap shows that the assumptions of the model can be defended for a particular scenario.

3.2.5 The Arbitrage-Free Value of a Eurobond

Denote the arbitrage-free price of the defaultable zero coupon bond with (net) maturity T by $F(T)$ and the standard normal distribution function evaluated at z by $N(z)$. Following the line of reasoning by Schönbucher (November 1997, p.23) the following closed-form solution can be proven:

Proposition:

$$F(T) \quad = \quad B(0,T)(1 - wQ^T) \tag{3.7}$$

$$Q^T \quad \stackrel{\text{def}}{=} \quad 1 - N\left(\frac{k - \frac{\Xi}{2}}{\sqrt[2]{\Xi}}\right) + \exp(k)N\left(\frac{-k - \frac{\Xi}{2}}{\sqrt[2]{\Xi}}\right)$$

$$k \quad \stackrel{\text{def}}{=} \quad \ln\frac{V_0}{\kappa(0,T)}$$

$$\Xi \quad \stackrel{\text{def}}{=} \quad \Xi_1 - \Xi_2$$

$$\Xi_1 \quad \stackrel{\text{def}}{=} \quad T\theta^2 + \frac{\sigma^2}{\lambda^2}\left[T - \frac{2}{\lambda}(1 - \exp(-\lambda T)) + \frac{1}{2\lambda}(1 - \exp(-2\lambda T))\right]$$

$$\Xi_2 \quad \stackrel{\text{def}}{=} \quad 2\rho\theta\frac{\sigma}{\lambda}\left[T - \frac{1}{\lambda}(1 - \exp(-\lambda T))\right]$$

Proof: At the outset some motivation is given. Part of it can be found in Musiela and Rutkowski (1997)[p.314ff]. The expected payoff of the defaultable zero bond under the equivalent martingale measure has to be calculated. The payoff is one if the value of the equity index stays above $\kappa(t, T)$ until T. The payoff is $1 - w$ if default occurs before T. Making use of the indicator function II that is one if default has occured and zero otherwise the payoff is $1 - w\mathrm{II}$. Using the universal risk neutral measure with the savings account as numeraire would result in calculation of the following expected value of the discounted payoff in order to obtain the arbitrage-free value:

$$E\left(\exp\left(-\int_0^T r(t)dt\right)(1 - w\mathrm{II})\right) \qquad (3.8)$$

Since $r(t)$ is stochastic the numeraire does not drop out of the expected value calculation. This makes the calculation of the integral difficult. Following the ideas presented in El-Karoui, Geman and Rochet (1995) a certain numeraire asset is convenient when calculating expected payoffs given interest rate uncertainty. The default-riskless zero bond maturing at time T is chosen as numeraire with T fixed. In the sequel an upperscript T denotes the use of the forward neutral measure. Making use of it the following expected value of the discounted payoff has to be calculated:

$$E^T\left(B(0,T)(1 - w\mathrm{II})\right) \qquad (3.9)$$

This time $B(0,T)$ can be put outside the expectation because it is known from today's discount curve. Hence, the fair value is given by:

$$F(T) = B(0,T)(1 - wQ^T) \qquad (3.10)$$

It will turn out that Q^T is the probability of the event that $\frac{V(t)}{B(t,T)}$ crosses K before T.

The Girsanov density for the change from the risk-neutral measure (denoted by $\mathcal{P}$ to the forward neutral measure $\mathcal{P}^T$ is:

$$\frac{d\mathcal{P}^T}{d\mathcal{P}} = \frac{\exp\left(-\int_0^T r(t)dt\right)}{B(0,T)} \qquad (3.11)$$

What remains to be done is calculation of Q^T. A time change is applied in order to cope with time-dependent volatility. For another use of this technique to a similar finite time horizon problem see Schmidt (1997). With the motivation in mind the actual calculations are performed now:

The dynamics of the equity index under the risk-neutral measure are governed by the following SDE and initial condition $V(0) = V_0$:

$$\frac{dV(t)}{V(t)} = r(t)dt + \rho\theta dw_1 + \theta\sqrt[2]{1 - \rho^2}dw_2 \qquad (3.12)$$

In order to obtain the new dynamics under the forward-neutral measure replace the stochastic differentials according to Musiela and Rutkowski (1997, p.466, Eq. (B.26)). The relation between the differentials under the two measures is as follows:

$$dw_1^T = dw_1 - \nu(t, T)dt \qquad (3.13)$$

$$dw_2^T = dw_2 \qquad (3.14)$$

The last equation is due to the fact that the differentials dw_i are independent and that the numeraire is driven by dw_1 alone. Inserting the last two equations yields the new dynamics for V:

$$\frac{dV(t)}{V(t)} = \left[r(t) + \nu(t, T)\rho\theta\right]dt + \rho\theta dw_1^T + \theta\sqrt[2]{1 - \rho^2}dw_2^T \qquad (3.15)$$

Note that the dynamics of the default riskless zero bond have changed as well:

$$\frac{dB(t, T)}{B(t, T)} = \left[r(t) + \nu(t, T)^2\right]dt + \nu(t, T)dw_1^T \qquad (3.16)$$

Expressing the index value $V(t)$ in units of the new numeraire gives rise to the definition of $\tilde{V}(t)$:

$$\tilde{V}(t) \stackrel{\text{def}}{=} \frac{V(t)}{B(t, T)} \qquad (3.17)$$

The dynamics of $\tilde{V}(t)$ result from an application of Ito's formula Musiela and Rutkowski (1997)[p.463] to the last two SDEs.

$$\frac{d\tilde{V}(t)}{\tilde{V}(t)} = [\rho\theta - \nu(t,T)]\, dw_1^T + \theta\sqrt[2]{1 - \rho^2}dw_2^T \qquad (3.18)$$

Because there is no drift in the last SDE, the T-forward measure is a martingale measure needed for pricing purposes. By construction prices relative to the zero bond $B(t,T)$ are martingales. Because the volatility at time t of a zero coupon bond maturing at time T is time-variant, $\tilde{V}(t)$ has time-dependent volatility. Before coping with this time-dependence a simplification of notation is introduced. A Brownian motion $\hat{w}$ is constructed from the two Brownian motions 1 and 2 that preserves the original volatility. Define a new deterministic function $\sigma(t)$:

$$\sigma(t) \stackrel{\text{def}}{=} \sqrt[2]{\theta^2 - 2\rho\theta\nu(t,T) + \nu(t,T)^2} \qquad (3.19)$$

Note the difference between the (short rate volatility) parameter σ and the function $\sigma(t)$. This defines the following martingale $M(t)$:

$$dM(t) \stackrel{\text{def}}{=} \sigma(t)d\hat{w} \qquad (3.20)$$

The quadratic variation of this martingale on the interval $[0,t]$ is denoted using the notation of Karatzas and Shreve (1997):

$$\langle M \rangle_t = \int_0^t \sigma(s)^2 ds \qquad (3.21)$$

Note that this is a Riemann integral. It has the property $\int_0^t \sigma(s)^2 ds < \infty, \forall t \geq 0$. Hence, theorem 4.8 of Karatzas and Shreve (1997)[p.176] is applicable and states that there is a new Brownian motion W with the property:

$$\tilde{V}(t) = \int_0^t \tilde{V}(s)dM(s) = \int_0^{\langle M \rangle_t} Y(u)dW(u) = Y(\langle M \rangle_t) \qquad (3.22)$$

The equation reveals, that $Y(u)$ satisfies the following SDE:

$$\frac{dY(u)}{Y(u)} = dW \tag{3.23}$$

Hence, $Y(u)$ follows a geometric Brownian motion with time measured on a new scale. In order to use this new scale it should be remembered that there is no default as long as:

$$V(t) > B(t,T)K, \forall t \le T \tag{3.24}$$

This is equivalent to the condition:

$$\frac{V(t)}{B(t,T)} > K, \forall t \le T \tag{3.25}$$

But the left-hand side is $\tilde{V}(t)$. This finding explains the specification of the time-variant barrier. With respect to the numeraire $B(t,T)$ the condition is as follows:

$$\tilde{V}(t) > K, \forall t \le T \tag{3.26}$$

Under the T-forward measure the barrier is no longer time-variant! Expressing the condition in terms of Y and the new time-scale it reads:

$$Y(u) > K, \forall u \le \langle M \rangle_T \tag{3.27}$$

With respect to a Brownian motion with drift starting at zero, the probability of the process staying above a barrier has been calculated already. Hence, dividing by the initial value of $Y(0) = Y_0$ and taking natural logarithms on both sides of the inequality yields the equivalent condition:

$$\ln \frac{Y(u)}{Y_0} > \ln \frac{K}{Y_0}, \forall u \le \langle M \rangle_T \tag{3.28}$$

Usage of corollary B.3.4 of Musiela and Rutkowski (1997)[p.470] it turns out that the default probability under the T-forward measure is as stated in the proposition. Multiplying the expected payoff with $B(0,T)$ yields the fair value. *q.e.d.*

Note the negative effect of the correlation parameter ρ. The intuition is as follows: Because the probability Q is calculated under the forward-neutral measure the decisive size is the relative price $\tilde{V}(t)$. Hence, its relative volatility counts. If $\rho = 1$ the numerator and denominator of $\tilde{V}(t)$ move randomly in the same direction thus diminishing the volatility of the fraction. This extreme case does not imply that the volatility is zero since both assets are affected differently by randomness.

Next another correlation is considered: The interdependence between the defaultable zero bond and the equity index. In the sequel it will turn out that the first partial derivative with respect to V_0 may be of interest. Denoting the standard normal density by n(.) this partial derivative is as follows:

$$\frac{\partial F(T)}{\partial V_0} = \frac{B(0,T)w}{V_0}\left[\Psi_1 + \Psi_2 + \Psi_3\right] \qquad (3.29)$$

$$\Psi_1 \overset{\text{def}}{=} \frac{1}{\sqrt[2]{\Xi}} n\left(\frac{k - \frac{\Xi}{2}}{\sqrt[2]{\Xi}}\right)$$

$$\Psi_2 \overset{\text{def}}{=} -\exp(k) N\left(\frac{-k - \frac{\Xi}{2}}{\sqrt[2]{\Xi}}\right)$$

$$\Psi_3 \overset{\text{def}}{=} \frac{1}{\sqrt[2]{\Xi}} \exp(k) n\left(\frac{-k - \frac{\Xi}{2}}{\sqrt[2]{\Xi}}\right)$$

Close inspection of the formula shows that the arbitrage-free value of a defaultable zero bond depends non-linearly on the value of the assets of the issuer. If there is a long distance between the value of the assets and the trigger level, the influence of the value of the assets is rather small. But if this value approaches the trigger the influence increases in exponential manner. An immediate consequence is a caveat concerning the use of correlation measures when analyzing country risk. On the one-hand correlation measures only the linear dependence between variables. On the other hand the samples will be dominated by data from "quiet times". Thus the dependence is understated twofold when measured by correlation. It should be noted that the formula and its partial derivative were derived for time zero (full maturity T). It is left to the reader to replace T by $T-t$ in order to allow for the passage of time. For practical hedging this notationally more expensive version of the partial derivative is used.

3.2.6   Possible Applications

Formula (3.7) (p.45) has the following economic meaning: Due to the assumptions of frictionless markets for the currency and equity index of the borrowing sovereign and the default-riskless zero coupon bonds denominated in the lending currency it is possible to build a portfolio consisting of a long position in the foreign equity index and the domestic zero bond, that has the same value as the Eurobond at any point of time between now and maturity. An immediate consequence is that a long position in the Eurobond can be hedged by being short a certain amount of the foreign equity index and the domestic zero bond. The latter position hedges the risk arising from the default-free term structure, the former the default risk.

This model allows also for the evaluation of a portfolio of Eurobonds issued by different sovereigns but denominated in one currency, e.g. DM. Not only correlation between the foreign equity index expressed in the currency of denomination and the default-riskless term structure is captured, but also the correlations between the equity indices and, hence, the so-called default correlation between different issuers as it derives from the correlations of the indices.

The above formula (3.7) (p.45) can be used to determine those parameters that fit best to the market prices of Eurobonds. Having extracted this information from the market place, it can be used for pricing new issues or to analyze a portfolio of Eurobonds incorporating correlation. This default correlation arises from the correlation of the equity indices and currencies.

Although the formula is given for a defaultable zero coupon bond in the sequel it is applied to coupon bonds as well. Thinking of a coupon bond as a portfolio of zero bonds one might remark that - given K in formula (3.6) (p.44) - the effective trigger level $\kappa(t, T)$ varies from payment date to payment date. A coupon due at time T_1 defaults if $\kappa(t, T_1)$ is hit before, whereas a coupon due at time T_2 defaults if $\kappa(t, T_2)$ is touched. Assume that T_1 is before (smaller than) T_2. Because the discount factor B(t,T) is bigger for T_1 than for T_2, $\kappa(t, T_1)$ is bigger than $\kappa(t, T_2)$ as can be seen by formula (3.6) (p.44). If one is aware of the fact that not the nominal value of outstanding debt but its present value is the decisive figure, it is no surprise that the assets of the issuer may turn out to be insufficient to service debt in the near future T_1 but sufficient to pay the coupon at a more remote time T_2. Nevertheless this handling of defaultable coupon bonds raises doubts and is therefore an area for future research.

Note that this handling of coupon bonds is in contrast to the default mechanism of Longstaff and Schwartz (1995). In their model all outstanding payments default if the first coupon defaults. However, it should be noted that in the case of "defaulting" sovereign debtors the outstanding payments are negotiated payment by payment.

Finally, possible applications in risk management are imaginable, but should be pursued with care as has been pointed out by Duffie and Pan (1997)[p.10]: "Derivatives pricing models are based on the idea ... that the price of a security is the expected cash flow paid by the security discounted at the risk-free interest rate. The fact that this risk-neutral pricing approach is consistent with derivatives pricing in efficient capital markets does not mean that investors are risk-neutral. Indeed the actual risk represented by a position typically differs from that represented in risk-neutral models (see Harrison and Kreps [1979]). For purposes of measuring value at risk at short time horizons such as a few days or weeks, however, the distinction between risk-neutral and actual price behavior turns out to be negligible for most markets. This means that one can draw a significant amount of information for risk measurement purposes from one's derivatives pricing models, *provided they are correct*. This proviso is checked in the sequel.

3.2.7 Determination of Parameters

Three sets can be distinguished:

Default related parameters

The current level of the issuer's equity index and currency have to be determined. Multiplication of both market figures yields V_0 as it appears in formula (3.7) (p.45). The volatility of $V(t)$, i.e. θ, has to be figured out as well. This can be estimated from historical time series of $V(t)$ or extracted from the equity- and currency derivatives markets.

The write down w and the nominal trigger level K have to be determined by sound economic reasoning or have to be extracted from Eurobond prices during a specified calibration period. Longstaff and Schwartz (1995)[p.794] propose to use the write down as a measure of the seniority of the bond. It can be expected that the default related parameters vary among issuers. Thus the objections by Hajivassiliou (1987), that the stochastics differ countrywise, are taken into account.

Term structure related parameters

The two parameters σ and λ may be extracted from the market for interest rate derivatives or estimated. These parameters have to be determined per currency of denomination.

Link parameter

This parameter is denoted by ρ. It links the stochastics of the default-free term structure to the default related dynamics of the equity index. It has to be determined per issuer using again an "imply-out" approach or estimation techniques.

3.3 Description of Data and Parameter Setting

First, the Eurobonds investigated are presented and their optional features are discussed. Then the default-related data from the equity and currency markets is introduced. Finally, construction of the default-free term structure and its dynamics are sketched. All data is either from DATASTREAM or a proprietary database of WestLB.

3.3.1 DM-Eurobonds under Consideration

The selection of the DM-Eurobonds is driven by aspects of data availability. The focus on emerging economies is due to the recent developments in financial markets.

All these sovereign issuers are considered, where good time series for the exchange rate and the equity index and the Eurobonds are available on a daily basis. All bonds considered have maturities up to 10 years. Pricing bonds with longer maturities is critical, because there are few default riskless bonds with such maturities. Hence, the discount function becomes a critical input for such maturities.

The following table describes the DM-Eurobonds investigated.

Table I
DM-Eurobonds under Consideration

Sovereign	Year of Issue	Coupon	Maturity Date
ARGENTINA	1995	$10\frac{1}{2}$	14/11/02
ARGENTINA	1995	$9\frac{1}{4}$	29/08/00
ARGENTINA	1996	$10\frac{1}{4}$	06/02/03
ARGENTINA	1996	7	20/05/99
ARGENTINA	1996	$8\frac{1}{2}$	23/02/05
ARGENTINA	1996	9	19/09/03
ARGENTINA	1997	7	18/03/04
BRAZIL	1997	8	26/02/07
MEXICO	1995	$9\frac{3}{8}$	02/11/00
MEXICO	1996	$10\frac{3}{8}$	29/01/03
POLAND	1996	$6\frac{1}{8}$	31/07/01
RUSSIA	1997	9	25/03/04

The offering circulars for the above issues reveal that there are at least three of the following seven options on the side of the investor (lender):

If any of the following events occurs and is continuing, the holder of the bond may declare such bond immediately due and payable together with accrued interest thereon:

Table II
Events

Non-Payment of Principal
Non-Payment of Interest
Breach of Other Obligations
Cross-Default
Moratorium
Contested Validity
Loss of IMF-Membership

These options are "written" (i.e. granted) by the issuer herself. Hence, there is strong negative correlation between the payoff of the options and the financial condition of the option-writer. When the option is "in the money" (i.e. has value), the issuer is "out of money" and vice versa. Judged from the perspective of the investor (lender) the value of these options is negligible. With respect to the exercise of these options it should be kept in mind that it is very difficult to enforce payment of the respective payoffs. Summing up all DM-Eurobonds are treated as being straight without optionalities. All have bullet repayment of principal and yearly coupons in DM.

3.3.2 Equity Indices and Currencies

This section gives the names of the equity indices used and reports the volatilities and correlations needed in formula (3.7) (p.45). The first trading day that is represented in the data is the 21st of April 1997. Including this date the data comprise 174 trading days in the year 1997. This nearly three-quarter period of 1997 is used as a so-called calibration period. The start of the calibration period is due to the Russian DM-Eurobond, that was issued in April 1997.

The meaning of the calibration period is as follows: In general parameters are determined with respect to this time interval and used for pricing in the following years. Statistically speaking the model is fine-tuned "in the sample 1997" and its performance is judged "out of the sample" in 1998. In some cases the equity indices expressed in the currency DM have been scaled down by a fixed factor as is shown in the following table. Note that all figures are in decimal dimensions, e.g. a volatility of 20 is 0.2.

In order to give detailed information on the issuers additional notation is introduced: The equity index of the borrowers country in its domestic currency is denoted by I and has a volatility of ϑ_I. The price of one unit currency of the issuer expressed in units of country L's currency is denoted by X and has a volatility of ϑ_X. The continuous returns of X and I have a correlation that is denoted by ϱ_{XI}.

Although the volatility of the DM-Index can be calculated from the preceding columns it is given for the reader's convenience. Note that only this figure enters formula (3.7) (p.45). All figures are rounded to two decimals. All except the penultimate column contain volatility-figures.

Table III
Issuer Related Historical Volatility

Sovereign	Index	Currency	Correlation	DM-Index
Equity Index and Scaling	ϑ_I	ϑ_X	ϱ_{XI}	θ
ARGENTINA *(Merval)*	0.36	0.1	0.31	0.41
BRAZIL *(Bovespa/100)*	0.54	0.1	0.23	0.57
MEXICO *(IPC)*	0.32	0.16	0.65	0.44
POLAND *(WIG/100)*	0.28	0.13	0.23	0.33
RUSSIA *(RTS)*	0.56	0.1	0.01	0.57

The same level of currency volatility for Argentina, Brazil, and Russia is no surprise since all three currencies were pegged to the US-Dollar during 1997. Factually on the 17th of August 1998 (officially on the 3rd

of September 1998) the Rubel started floating and on 13th of January
1999 the Real did too. Because the period under consideration ends
with August 1998 the volatility parameters are not changed because the
sample of "floating Rubel" is too small in order to estimate a "new"
volatility.

3.3.3 Default-Free Term Structure and Correlation

Due to the term structure model being no-arbitrage a good discount
curve of the German default-riskless yield curve is needed and an adequate volatility function. For each trading day under consideration
more than 100 government bonds were used to calculate the discount
curve. The parameters of the volatility function were chosen in a way
that pricing of two REX-linked bonds with maturities 2000 and 2005 was
perfect during the calibration period from 21 April to 31 December 1997
($\sigma = 0.018$ and $\lambda = 0.046$). Because the default-riskless DM-short rate
$r(t)$ is not an observable quantity, the parameter ρ measuring correlation
between the the default-free term structure and the default related dynamics of the equity index cannot be determined empirically. Therefore
it is set to zero, although - theoretically - the short rate could be derived
from the discount curve. But this would introduce other problems: The
discount curve is built from the market prices of DM-goverment bonds.
It is well known that the short end of a discount curve built that way
should be considered with care. Hence, the derived short rate would
rather be a theoretical artefact than a real market rate.

3.3.4 Calibration of Default-Mechanism

Inspection of formula (3.7) (p.45) reveals that the theoretical value of a
defaultable bond is "Default-Riskless Bond times Adjustment" and that
the "Adjustment" is given by "One minus (Pseudo-) Default-Probability
(i.e. Q) times Write Down (i.e. w)". Hence, the "write down" enters the
formula in the same manner as the (pseudo-) default-probability. From
Moody's research Moody's (1998)[p.19] in the corporate bond market it
is known that the average magnitude of the write down is around 0.6.

Write Down

First, the available number of countries, that have defaulted since World
War II, is far too small in order to allow for statistically meaningful
estimation of recovery rates. A Moody's-like approach to recovery rates
of sovereign bonds is therefore impossible. But Moody's insights into
recovery rates of corporate bonds help to exclude certain dimensions of

the recovery rate. Because it is very difficult to enforce certain rights with respect to sovereign debtors it can be expected that the "write down" of Eurobonds issued by sovereigns is not smaller than that for corporates.

Judgemental "write down" values of 0.6 and 0.7 are chosen for the pricing of DM-Eurobonds. The lower value is assigned to an issuer if during 1997 the correlation between the market price of the DM-Eurobond and its (fictive) default-riskless counterpart is higher than 0.5. Otherwise $w = 0.7$ is chosen. This correlation is interpreted as a signal from the market concerning the severity of loss given default. In effect only the Russian Eurobond is priced using a write down of 70 percent. The remaining Eurobonds are valued with $w = 0.6$.

Default-Trigger

The 174 trading days in the year 1997 are used as a so-called calibration period. The level of the nominal default trigger K is chosen per issuer - not per bond - in a way that the average absolute pricing errors are minimal during the three last quarters of 1997. Note that any economic approach to derive K from the level of the issuer's liabilities in foreign currencies would cause serious problems, since it is not clear which fraction of it corresponds to the "assets" of the economy approximated by the equity index.

The resulting values for K are reported in the next table as well as the step size in the search procedure (bisection). The step size varies per issuer, because the equity indices are in different dimensions. Hence, the resulting trigger levels are rounded figures. The following table summarizes the results and reports the levels of the equity indices as of 21st April 97 for the reader's convenience.

Table IV
Default Related Parameters

Sovereign (Equity-Index)	DM-Index	DM-Trigger	Step Size in Search
	$V()$	K	
ARGENTINA (Merval)	1185.79	400	50
BRAZIL (Bovespa/100)	150.67	26	1
MEXICO (IPC)	810.79	250	50
POLAND (WIG/100)	90.21	20	1
RUSSIA (RTS)	87.19	17	1

3.4 Pricing Capability

3.4.1   Test Methodology

To judge upon the realism of formula (3.7) (p.45) nearly the same approach as the one by Bühler, Uhrig-Homburg, Walter and Weber (1999) (BUWW) [p.270] with respect to interest-rate options is applied: "First, valuation models within risk management systems must be capable of predicting future option prices if they are to correctly measure risk exposure. This capability is best evaluated by the ex ante predictability of a model. Therefore, we use the valuation quality of a model, not its ability to identify mispriced options, as the most important assessment criterion".

This methodology can be applied to DM-Eurobonds because they can be viewed as term structure derivatives. To provide a yardstick for measuring pricing quality the following quotation may be helpful. BUWW examine the empirical quality of the models by comparing model prices to market prices for the period from 1990 to 1993. From model to model the "average absolute pricing errors vary between 21 percent and 37 percent" Bühler et al. (1999) [p.292].

Keeping this in mind the test proceeds as follows: Beginning on the 21st of April 1997 daily prices are calculated using formula (3.7) (p.45) and compared to market prices. The inputs are described below.

3.4.2   Inputs for the Closed-Form Solution

The inputs that are updated daily are:

Default-Free Term Structure

The daily update of the default-free term structure, which is calculated from the market prices of DM-government bonds, influences the level of the default-riskless discount factor $B(0,T)$.

Equity Index

The current level of the issuer's equity index and currency are determined on a daily basis. Multiplication of both market figures yields V_0 as it appears in the formula.

All other inputs in formula (3.7) (p.45) are pegged to the levels reported in the tables III and IV.

3.4.3 Model versus Market Prices

Because any model is a simplification of reality the model prices will deviate from the market prices. The average absolute pricing error is calculated per issuer as follows: On each trading day the absolute deviation between model and market price per outstanding bond is calculated as a percentage. Thus, there is no leveling out of over- and underpricing. Then these daily percentage figures are averaged over all trading days (of the respective year) and bonds defining a unique average absolute pricing error per issuer and year. The number of bonds considered in averaging is reported in the second column. It should be noted that with respect to 1997 only the last three quarters are considered, whereas in 1998 the time series ends on the 31st of August thus covering nearly the first three quarters of 1998. The last two columns are percentage figures, e.g. during the first three quarters of 1998 the overall "Average Absolute Pricing Errors" with respect to seven Eurobonds issued by Argentina is 3.24 percent.

Table V
Average Absolute Pricing Errors

Issuer	No of Bonds	1997	1998
ARGENTINA (Merval)	7	1.50	3.24
BRAZIL (Bovespa/100)	1	0.77	1.33
MEXICO (IPC)	2	1.84	1.23
POLAND (WIG/100)	1	0.35	0.69
RUSSIA (RTS)	1	2.19	5.32

Of course, a more frequent update of the parameters $\sigma, \lambda, \theta, \rho, w$, and K would increase the match with market prices considerably. Note that these parameters were left unchanged during the time from April 1997 to August 1998. But this could induce the criticism of "data cooking" and would impede the workability of the approach. Finally, it should be emphasized that Eurobonds are less volatile than the interest rate derivatives investigated by Bühler et al. (1999). This should be kept in mind when comparing the error figures.

3.5 Hedging

The theoretical assumptions behind formula (3.7) (p.45) assure that there is a dynamic portfolio strategy consisting of certain positions in the default riskless zero bond and the foreign equity index that - when continuously adjusted - replicate the price dynamics of the DM-Eurobond.

Since this situation is not found in reality this section is devoted to *practical* hedging of a long position in a DM-Eurobond. Solely for illustrative purposes this hedging is presented in two parts that are sequentially discussed. Eventually these parts will be put together.

3.5.1 Static Part of Hedge

"Go short" the fictive default-riskless counterpart of the DM-Eurobond which has the same maturity and coupon. Note that in the case of no default of the DM-Eurobond the cash flows cancel. This strategy corresponds to lending "Deutsche Mark" to the sovereign and refinancing by the lender through the issue of bonds that are assumed to be default-riskless.

Hence, with respect to the available data the strategy reads: In April 1997 buy one DM-Eurobond (from the menu in Table I), refinance this long position by issuing a default-riskless bond with the same maturity and coupon. On the 31st of August 1998 sell the DM-Eurobond and buy a default-riskless bond by another default-free issuer in order to close the refinancing position. The following table contains the respective price differences from a cash-flow perspective. All figures are in "Deutsche Mark" (i.e. DM). The column "interest" results from the fact that the purchased DM-Eurobond has a market value less than the bond issued by the lender. The initial price difference is assumed to be invested in a default riskless zero bond maturing on the 31st of August 1998. Note that the coupons of the coupon bond positions cancel.

Table VI
Results of Static Part

Sovereign	DM-Eurobond	Refinancing	Interest	Maturity Date
ARGENTINA	-12.55	00.74	0.74	14/11/02
ARGENTINA	-09.70	04.59	0.31	29/08/00
ARGENTINA	-11.25	-00.09	0.76	06/02/03
ARGENTINA	-12.15	04.16	0.12	20/05/99
ARGENTINA	-04.80	-06.52	0.76	23/02/05
ARGENTINA	-06.50	-02.90	0.74	19/09/03
ARGENTINA	-07.15	-02.75	0.37	18/03/04
BRAZIL	-16.65	-11.08	0.87	26/02/07
MEXICO	-16.75	04.41	0.35	02/11/00
MEXICO	-21.90	00.09	0.71	29/01/03
POLAND	-00.60	-00.86	0.08	31/07/01
RUSSIA	-60.43	-04.08	1.04	25/03/04

3.5.2 Dynamic Part of Hedge

Since the purchased DM-Eurobond has a market value less than the bond issued by the lender there is a cash inflow to the lender. On the other hand in the case of default the lender has to service its own issue whereas the income from the DM-Eurobond is reduced - in nominal terms - according to the write down percentage w. In essence the cash inflow of the lender is the "received" premium of a "written" (issued, shorted) derivative. In terms of the model that is applied to the pricing of DM-Eurobonds the derivative corresponds to a long position in the equity index of the sovereign with limited upside potential. Hedging this position requires a counterposition, i.e. a short position. Therefore, the dynamic hedge works as follows:

"Go short" a certain amount Δ of the equity index and receive foreign currency. Exchange this for "Deutsche Mark" at the current exchange rate. This yields a cash inflow of ΔV_0. In theory this position is revised every instant. Practically, weekly revisions may seem adequate. Because the 31st of August 1998 is a Monday and the markets were very volatile in the preceeding week, it is assumed that on every Monday the Δ-position in the equity index is closed and built up again. This would be a waste of resources if the amount Δ was static. In fact it is not. Using the central insight of Black and Scholes (1973) the parameter Δ is given by the time-t dependent version of formula (3.29) (p.50). This depends on the actual level of the equity index V at time t. If the "Mondays" of the time under consideration are indexed by $T_1, T_2, ..., T_i$ and the time varying amounts Δ also, the weekly cash flow results from the dynamic hedge can be given. For instance on the first Monday (i.e. 21st of August 1997) the index is sold (shorted) and on the second Monday (i.e. 28th of August 1997) it is bought giving rise to the cash flow result $\Delta(T_1)\left[V(T_1) - V(T_2)\right]$. In the sequel interest rate effects are neglected, since the cash positions are rather small. Nearly the same cash flows result from hedging with the future on the respective index. In this case, which is of more practical relevance, "going short the index" is replaced by "selling the future". Now, the cash flow occurs when closing the position by buying back the future.

$$
\begin{aligned}
CashFlowAtT_2 &= \Delta(T_1)\left[V(T_1) - V(T_2)\right] \qquad (3.30)\\
CashFlowAtT_3 &= \Delta(T_2)\left[V(T_2) - V(T_3)\right]\\
\vdots\ &=\ \vdots\\
CashFlowAtT_i &= \Delta(T_{i-1})\left[V(T_{i-1}) - V(T_i)\right]
\end{aligned}
$$

Last but not the least it should be mentioned that the Austrian Futures and Options Exchange and the Chicago Mercantile Exchange offer the respective futures. One interesting alternative to selling the future on the index is the issue of a certain amount of a so-called index-certificate. But in contrast to the futures markets the adjustment of the position is dependent on investors' sentiment. Hence, selling the certificate on the index of the borrower is easy when the "delta" is small, i.e. when the need for hedging is negligible. But if things get worse with the economic condition of the borrowing sovereign investors will stay away from investing in (more) index-certificates. In contrast, the presence of speculators on the futures exchanges guarantees liquidity. The degree of liquidity can be measured by the so-called open interest. To give an impression table VII displays open interest on the 30th of March 1999 (source Bloomberg):

Table VII
Liquidity Of Some Futures

Sovereign	Open Interest
(Exchange)	
BRAZIL *(Bolsa de MeF)*	21806
POLAND *(AFO)*	1909
RUSSIA *(AFO)*	3240

3.5.3   Evaluation of the Hedging Strategy

Summing the weekly cash flows yields the result of the dynamic hedge which are given in the second column per Eurobond. For the reader's convenience the net results from the static hedge are repeated in the third column. Finally, both results are summed up and reported in the last column. Thus the results from the practical hedge position are evident considering both parts in sum.

Table VIII
Results of Two-Part Practical Hedge

Sovereign	Dynamic	Net Static	Sum	Maturity Date
ARGENTINA	17.06	-11.07	05.99	14/11/02
ARGENTINA	13.58	-04.80	08.78	29/08/00
ARGENTINA	17.12	-10.58	06.54	06/02/03
ARGENTINA	03.39	-07.78	-04.39	20/05/99
ARGENTINA	16.18	-10.56	05.62	23/02/05
ARGENTINA	16.64	-08.66	07.98	19/09/03
ARGENTINA	15.54	-09.53	06.01	18/03/04
BRAZIL	03.99	-26.86	-22.87	26/02/07
MEXICO	02.27	-11.99	-09.72	02/11/00
MEXICO	05.81	-21.10	-15.29	29/01/03
POLAND	00.31	-01.38	-01.07	31/07/01
RUSSIA	48.70	-63.47	-14.77	25/03/04

The table highlights the difference between the static approach on a stand-alone basis and in combination with the dynamic add-on. It may be resumed that the add-on of the dynamic part lightens the losses from lending to "emerging" sovereigns.

The weekly rebalancing might appear cumbersome. A static alternative to the dynamic hedge could be the purchase of a certain amount of some kind of put option on the equity index V. This would have the advantage that the holder of the put would profit from an increase in the volatility in V and not only from the market moves of V. In effect, this purchase of a put would delegate the dynamic hegding to the writer of the put. Because this put can be produced only on the OTC-market for exotic derivatives it will be costly. Therefore this option is not discussed in detail.

The case of Russia is of special interest because it is the only borrower where a default in the sense of the model occured. The nominal trigger level K was set to DM 17 through calibration with respect to 1997. On the 27th of August 1998 the Russian equity index fell to DM 14.45 from DM 17.53 the day before. As mentioned in the introduction of the model this move in the equity and currency markets did not trigger default for all outstanding payments of coupon and principal of the Russian DM-Eurobond. In fact, only the "near" coupons in the years from 1999 to 2002 defaulted, because the effective trigger $\kappa(t, T)$ is the discounted value of K. Riskless discounting with the discount curve from the 27th of August 1998 yields the values in the table.

Table IX
Effective Trigger Values

Payment Date	Respective Trigger
25th March 1999	16.67
25th March 2000	16.07
25th March 2001	15.47
25th March 2002	14.86
25th March 2003	14.22
25th March 2004	13.59

3.6   Management of a Portfolio

Recently the analysis of a pool of loans has been performed by McAllister and Mingo (1996). An analogous approach is applied to a Eurobond portfolio. Optimal conditions for a portfolio in general are derived and discussed. Finally, the simulated Eurobond portfolio is analyzed.

3.6.1   Set Up of the Monte Carlo Approach

In the structural No-Arbitrage approach presented above "the financial condition of borrower i is represented by" McAllister and Mingo (1996) [p.1387] the equity index expressed in the currency of the lender (e.g. DM). This variable has the advantage of being observable and tradeable. In the following analysis i is from the set of sovereign issuers introduced above. A Monte Carlo approach is applied using 10.000 pathes. In addition term structure uncertainty is simulated as well. Keeping in mind the proviso by Duffie and Pan - that "for purposes of measuring value at risk at short time horizons such as a few days or weeks, .. the distinction between risk-neutral and actual price behavior turns out to be negligible" Duffie and Pan (1997) - the time horizon for the simulation is two months ($=2/12=0.167$ units of a year). In the presence of term structure uncertainty the so-called short rate $r(t)$ has to be simulated stepwise. Choosing 30 steps per path each step corresponds to two calendar days. Because the equity index triggers default it has to be monitored very closely. It is obvious that the monitoring frequency of one simulated move (step) per two days can be improved. Instead of increasing the number of steps the ideas of Andersen and Brotherton-Ratcliffe are applied making the Monte Carlo technique a little more "exact" Andersen and Brotherton-Ratcliffe (1996). The quality of the random numbers is such that the first and second moment and the correlation structure are met exactly.

In order to balance the impact of each sovereign on the portfolio one zero coupon DM-Eurobond per sovereign with uniform initial maturity of three years is in the portfolio. Note that lending to sovereigns is no short term business.

The simulation makes use of the parameters that were described above. In addition the correlation matrix is estimated using the available continuous daily returns during the year 1997. Thus the volatility and correlation parameters are from the same "calibration period". Although the model presumes stability of the parameters they may be unstable in time.

Table X
DM-Index-Correlations

Sovereign	ARGENTINA	BRAZIL	MEXICO	POLAND	RUSSIA
ARGENTINA	1	0.77	0.82	0.29	0.29
BRAZIL	0.77	1	0.71	0.31	0.27
MEXICO	0.82	0.71	1	0.25	0.18
POLAND	0.29	0.31	0.25	1	0.58
RUSSIA	0.29	0.27	0.18	0.58	1

The following table reports the levels of the equity indices as of 5th January 98. These levels are used as start values in the Monte Carlo approach.

Table XI
Start Values for DM-Indices

Sovereign (Equity-Index)	DM-Index
ARGENTINA (Merval)	1243.11
BRAZIL (Bovespa/100)	173.43
MEXICO (IPC)	1181.09
POLAND (WIG/100)	76.97
RUSSIA (RTS)	126.05

The simulation evolves under the so-called risk-neutral measure with the riskless savings account as numeraire. The initial term structure is assumed flat at five percent in order to abstract from the real term structure. This gives rise to the following start values of the three-year defaultable zero bonds. Note that the default riskless three-year zero bond has a value of 86.38.

Table XII
Start Values for Zero Bonds

Sovereign	Theoretical Value
ARGENTINA	79.60
BRAZIL	81.66
MEXICO	83.39
POLAND	85.41
RUSSIA	81.60

Hence, the start value of the DM-Eurobond portfolio is 411.66. For each zero coupon DM-Eurobond in the portfolio there are two possible outcomes at the end of each path (i.e. in two months): Either default has occured causing a transformation of the Eurobond position to $(1 - w)$ units of a default-riskless zero bond with 34 months time to maturity and a value according to the level of the riskless term structure captured through the "one" factor short rate (Note: The shape of the term structure changes as well in the Extended Vasicek Model); or the defaultable Eurobond is still alive and has a value according to the level of the equity index and the short rate.

3.6.2   Optimality Condition

Traditional management of a portfolio uses "Value at Risk" as risk measure and compares investment alternatives according to their expected return relative to their contribution to the "Value at Risk" (see for instance Lehrbass (1999)). There is the proposal to replace VaR by a 'coherent' (in the sense of ADEH Artzner, Delbaen, Eber and Heath (1997)) risk measure. ADEH call a risk measure coherent if it satisfies the four relations of sub-additivity, homogeneity, monotonicity and the risk-free condition and show that VaR is not coherent. In contrast, the so-called "shortfall" risk measure is Embrechts, Klüppelberg and Mikosch (1997) [p.295]. This measure is specified below where the optimality condition is applied. In this subsection sufficient optimality conditions using the shortfall risk measure are derived by applying non-linear programming to the decision problem of a bank. Details concerning non-linear programming can be found in standard textbooks such as Chiang (1984). Imagine a bank that has at its disposal an investment amount M that earns the return R. To keep things tractable capital letters signify absolute amounts, whereas corresponding small letters denote relative (decimal) figures. Making use of this rule the percentage return of the bank is given by $r = R/M$. It should be noted that the short rate is denoted by $r(t)$, whereas r without the time argument signifies the percentage return of the bank over a specific time horizon. The amount of money invested in alternative one (two) is signified by A (B). Note that for a financial institution the restriction $A + B = M$ can be ignored completely. The percentage return of an alternative is denoted by r_A respectively r_B. With respect to the future these figures are uncertain. Hence, the concept of expected or anticipated returns has to be introduced. Let $E(r_A)$ respectively $E(r_B)$ denote the anticipated return of the alternatives under the real-world probability measure. It is assumed that there is a (differentiable) non-linear function called $SF(A, B)$ that maps the amounts invested into the alternatives (i.e. A and B) to the shortfall-figure of the portfolio. The decision-problem is to maximize the expected return $E(R) = E(r_A)A + E(r_B)B$ given a target level L of the banks shortfall. This restriction can be expressed as $SF(A, B) = L$. Because shortfall is a coherent risk measure it is convex! Hence, it turns out that the sufficient conditions for an optimal decision are as follows, where γ is the shadow price of the SF-resource:

$$E(r_A) = \gamma \frac{\partial SF()}{\partial A} \tag{3.31}$$

$$E(r_B) = \gamma \frac{\partial SF()}{\partial B} \tag{3.32}$$

$$SF(A, B) = L \tag{3.33}$$

Assuming that the bank does not waste the SF-resource only conditions 3.31 and 3.32 are of interest. Division of these by γ and equating yields the following sufficient optimality condition:

$$\frac{E(r_A)}{\frac{\partial SF()}{\partial A}} = \frac{E(r_B)}{\frac{\partial SF()}{\partial B}} \tag{3.34}$$

Condition 3.34 has the interpretation: Invest in such a way that the expected percentage return divided by the increase of the portfolio-shortfall is the same for all alternatives. The advantage of the partial derivative view is to highlight the portfolio effect. There is a disadvantage of the partial derivative view: Marginal, i.e. stepwise change of A and B is practically impossible, because in banking you usually face "take it or leave it" alternatives. Therefore the differentials in the optimality condition 3.34 are replaced by differences and transform the terms as follows:

$$\frac{E(r_A)}{\frac{\Delta SF()}{\Delta A}} = \frac{E(r_B)}{\frac{\Delta SF()}{\Delta B}} \qquad (3.35)$$

There is one more practical advantage to be mentioned: No knowledge of the shadow price γ is required.

But a severe limitation has to be pointed out. The decision problem leading to the optimality condition 3.35 is formulated under the real-world measure whereas pricing and the Monte Carlo simulation make use of other probability measures (forward- resp. risk-neutral). Therefore the expected return calculated under these "other" measures is already known by construction. Nevertheless keeping in mind the proviso by Duffie and Pan (1997) a ranking of the alternatives according to their riskiness is possible. Therefore condition 3.35 is rewritten as follows:

$$\frac{\frac{\Delta SF()}{\Delta B}}{\frac{\Delta SF()}{\Delta A}} = \frac{E(r_B)}{E(r_A)} \qquad (3.36)$$

This gives a "rough" guideline for ranking required expected returns under the real world measure. If an alternative has twice the risk contribution as another one, it should have twice the expected return.

3.6.3   Application of the Optimality Condition

Because the risk measure shortfall is sometimes called "beyond-VAR" the one-percent "VAR" of the initial DM-Eurobond portfolio is chosen as threshold level. The so-called "net worth" Artzner et al. (1997) of the portfolio is defined as "initial investment (e.g. 411.66 DM) minus the value of the portfolio after two month". The net worth figure is denoted by x. Hence, shortfall is given by the conditional expected value $E(-x|x \leq VAR)$. The VAR-figure from the simulation is -26.46 DM. Embrechts et al. (1997)[p.288] advocate "the use of the generalised Pareto distribution as a natural parametric model" for measuring shortfall. More specifically they suggest that below the threshold "standard techniques can be used" and above the threshold extreme value theory Embrechts et al. (1997)[p.359]. They point out that the generalised Pareto distribution "appears as the limit distribution of scaled excesses over high thresholds" Embrechts et al. (1997)[p.164] and provide information on "Fitting the GPD" Embrechts et al. (1997)[p.352]. A thorough application of extreme value theory is beyond the scope of this paper. Another limitation is the use of the risk neutral measure for calculation of $E(-x|x \leq VAR)$. In the base scenario the portfolio consists of one unit DM-Eurobond per sovereign. This leads to a shortfall of 0.304482 DM. The ratios $\frac{\Delta SF()}{\Delta A}$ are calculated as follows: The denominator signifies the position change in the portfolio in value terms, i.e. DM. Comparative statics are performed using a reduction of DM 10 in a specific DM-Eurobond position, i.e. $\Delta A = -10 DM$. The negative sign is due to A being the amount invested. For instance, in the case of "Argentina" instead of holding one unit of the zero bond only 0.874376159 units of the DM-Eurobond are in the portfolio. The next table contains the fraction $\frac{\Delta SF()}{\Delta A}$ per sovereign.

Table XIII
Shortfall Analysis

Sovereign	$\Delta SF()/\Delta A$
ARGENTINA	0.00747
BRAZIL	0.00740
MEXICO	0.00555
POLAND	0.00473
RUSSIA	0.00683

Note that a reduction of any component in the portfolio decreases the shortfall. This decrease differs among the sovereigns considered. As a general rule: The riskier (judged upon by the initial bond price) a sovereign, the more reaction in shortfall. Beside this individual - perspective rule is a correlation-based rule: The more correlated the "financial condition of the borrower" to the portfolio, the higher is the reaction in shortfall. This explains the ranking between Russia and Brazil. Although the bond prices are very similar (81.60 resp. 81.66) the impact of Brazil is higher due to the concentration of the portfolio in South-America.

It should be pointed out that the original shortfall was measured using the outcomes of one percent of the simulated pathes. Reducing the investment in a certain country as was done in the comparative statics decreases the VAR-figure and leads to a smaller number of outcomes used in measuring the shortfall below the original VAR-figure.

3.6.4   Modification of the Optimality Condition

If the risk measure shortfall is replaced by the "incoherent" and non-convex risk measure "Value at Risk" the following analysis emerges. Note that in all cases considered the VAR-figure decreased and that the decimal precision has not been changed.

Table XIV
Value at Risk Analysis

Sovereign	$\Delta VAR()/\Delta A$
ARGENTINA	0.12000
BRAZIL	0.12000
MEXICO	0.08000
POLAND	0.04000
RUSSIA	0.08000

Note that the optimality condition is no longer sufficient. It is merely necessary. The correlation effects are evident again. Because the portfolio is concentrated in South America, the risk contribution of Russia is relatively small.

3.7   Summary and Outlook

A simple model for the pricing of DM-Eurobonds has been developed. It is simple insofar as a continuous observable process is assumed to be the default-triggering factor. Hence, only few parameters have to be determined. Due to the lack of data on defaults even the determination of the parameters of a model such simple poses severe problems that have been overcome by judgement. The pricing capability has been evaluated and practical hedging discussed. The ranking of the riskiness varies with the perspective chosen. Because this point is important the next table summarizes the three perspectives (individual vs. VAR vs. SF):

Table XV
Different Perspectives of Riskiness

Sovereign	Value of ZeroBond	$\Delta VAR()/\Delta A$	$\Delta SF()/\Delta A$
ARGENTINA	79.60	0.12000	0.00747
BRAZIL	81.66	0.12000	0.00740
MEXICO	83.39	0.08000	0.00555
POLAND	85.41	0.04000	0.00473
RUSSIA	81.60	0.08000	0.00683

The dependence on the base portfolio has been pointed out. The example has revealed a sharper focus of the shortfall risk measure in comparison to traditional VAR.

Knowing what the adequate risk measure is (e.g. shortfall or VAR?) the optimality condition can be used to "search for" (*the* resp. *an*) optimal allocation of country limits. If shortfall is the choice it certainly is "a must" to devote more time on extreme value theory. Besides, there are a lot of outstanding refinements on the model itself and the parametrization. Future research is looked forward to.

Bibliography

Andersen, L. and Brotherton-Ratcliffe, R. (1996). Exact exotics, *RISK* **9(10)**: 85–89.

Artzner, P., Delbaen, F., Eber, J.-M. and Heath, D. (1997). Thinking coherently, *RISK* **10(11)**: 68–71.

Black, F. and Scholes, M. (1973). The pricing of options and corporate liabilities, *Journal of Political Economy* **81**: 637–654.

Branson, W. H. (1979). *Macroeconomic Theory and Policy*, number 2nd edition, Harper and Row, New York.

Bühler, W., Uhrig-Homburg, M., Walter, U. and Weber, T. (1999). An empirical comparison of forward-rate and spot-rate models for valuing interest-rate options, *Journal of Finance* **54**: 269–305.

Chiang, A. C. (1984). *Fundamental Methods of Mathematical Economics*, McGraw-Hill, New York.

De Haan, J., Siermann, C. L. J. and Van Lubek, E. (1997). Political instability and country risk: new evidence, *Applied Economics Letters* **4**: 703–707.

Duffie, D. and Lando, D. (1997). Term structures of credit spreads with incomplete accounting information, *Technical report*, Graduate School of Business, Stanford University,.

Duffie, D. and Pan, J. (1997). An overwiew of value at risk, *Journal of Derivatives* **4**: 7–49.

El-Karoui, N., Geman, H. and Rochet, J.-C. (1995). Changes of numeraire, changes of probability measure and option pricing, *Journal of Applied Probability* **32**: 443–458.

Embrechts, P., Klüppelberg, C. and Mikosch, T. (1997). *Modelling Extremal Events*, Springer, Berlin.

Erb, C. B., Harvey, C. R. and Viskanta, T. E. (1995). Country risk and global equity selection., *Journal of Portfolio Management* **10**: 74–83.

Grossman, S. J. and Stiglitz, J. E. (1980). On the impossibility of informationally efficient markets, *American Economic Review* **70**: 393–408.

Hajivassiliou, V. A. (1987). The external debt repayments problems of ldc's., *Journal of Econometrics* **36**: 205–230.

Harrison, J. M. and Kreps, D. M. (1979). Martingales and arbitrage in multiperiod securities markets., *Journal of Economic Theory* **20**: 381–408.

Heath, D., Jarrow, R. and Morton, A. (1992). Bond pricing and the term structure of interest rates: A new methodology for contingent claims valuation, *Econometrica* **60**: 77–105.

Ho, T. S. and Lee, S.-B. (1986). Term structure movements and pricing interest rate contingent claims, *Journal of Finance* **41**: 1011–1028.

Hull, J. and White, A. (1990). Pricing interest-rate-derivative securities, *Review of Financial Studies* **3**: 573–592.

Kaminsky, G., Lizondo, S. and Reinhart, C. M. (July 1997). Leading indicators of currency crisis, *Working Paper 97/79*, International Monetary Fund.

Karatzas, I. and Shreve, S. (1997). *Brownian Motion and Stochastic Calculus*, Vol. 2nd edition, Springer, New York.

Lehrbass, F. B. (1999). Rethinking risk-adjusted returns, *Credit Risk Special Report - RISK* **12(4)**: 35–40.

Longstaff, F. and Schwartz, E. (1995). A simple approach to valuing risky fixed and floating rate debt, *Journal of Finance* **50**: 789–819.

McAllister, P. H. and Mingo, J. J. (1996). Bank capital requirements for securitized loan pools, *Journal of Banking and Finance* **20**: 1381–1405.

Merton, R. C. (1973). Theory of rational option pricing, *Bell Journal of Economics and Management Science* **4**: 141–183.

Moody's (1998). Historical default rates of corporate bond issuers, 1920-1997, *Technical report*, Moody's Special Comment.

Musiela, M. and Rutkowski, M. (1997). *Martingale Methods in Financial Modelling*, Springer, New York.

of the Basel Committee, R. M. G. (1999). Principles for the management of credit risk, *Technical report*, Basel Committee on Banking Supervision, Basel.

Schlögl, E. and Sommer, D. (1994). On short rate processes and their implications for term structure movements, *Technical report*, Basel Committee on Banking Supervision, University of Bonn.

Schmidt, W. M. (1997). On a general class of one-factor models for the term structure of interest rates, *Finance and Stochastics* 1: 3–24.

Schönbucher, P. J. (November 1997). The pricing of credit risk and credit risk derivatives, *Technical report*, Department of Statistics, Faculty of Economics, University of Bonn.

Somerville, R. A. and Taffler, R. J. (1995). Banker judgement versus formal forecasting models: The case of country risk assessment, *Journal of Banking and Finance* **19**: 281–297.

Vasicek, O. (1977). An equilibrium characterization of the term structure, *Journal of Financial Economics* **27**: 117–161.

Zhou, C. (March 1997). A jump-diffusion approach to modeling credit risk and valuing defaultable securities, *Technical report*, Federal Reserve Board, Washington.

4 Predicting Bank Failures in Transition: Lessons from the Czech Bank Crisis of the mid-Nineties

Jan Hanousek

4.1 Motivation

Almost all countries in Central and Eastern Europe have experienced turbulence in their banking and financial sectors. In the early 1990s, for example, Poland's banks experienced a crisis, followed in 1994-1996 by the failure of several small banks in the Czech Republic, and severe problems in Latvia in 1995 when four of its large banks fell. Bank regulators in transition countries have been searching for early warning signals that could be used to make bank supervision more efficient. Although several papers exist on modeling or predicting bank failure in mature market economies [See Looney, Wansley and Lane (1989), Lane, Looney and Wansley (1990), Barber, Chang and Thurston (1996), Hwang, Lee and Liaw (1997); among others], there are several problems connected with the direct use of these models in transition economies. First, these models depend almost exclusively on economic conditions and balance-sheet data based on accounting standards that are conceptually and significantly different from those in transition economies. For example, most transition economies still use accounting procedures carried over from central planning that reflect production rather than profit. Moreover, unlike stable economies, the transition period is typified by high uncertainty, the lack of standard bank behavior, and other problems carried forward from the communist era that only worsen the situation. Finally, the vast majority of banks in transition economies have a very short history, with balance sheets seldom if ever scrutinized by prudent auditors.

The transition from a centrally planned to a market oriented economy is a complicated process with significant implications for the banking sector. In particular, it requires a separation of the central bank from commercial banks, the creation of a number of commercial banks from the single state bank, and the granting of licenses to new banks. The main problem facing the newly-emerging banking sector is the lack of expertise in credit evaluation under free-market conditions and the tendency to favor borrowers with fixed assets as collateral. Earning potential is largely ignored due to lack of ability to evaluate and proper accounting techniques [see EBRD (1996/7)]. Given the problematic inherited loan book and the high uncertainty typical of transition economies, banks are faced with extremely high risk. This situation is further worsened by the lack of three important factors: 1) experienced bank management; 2) market-enforced performance; and 3) standard measures to enforce the spreading of risk and prudent bank behavior

Although it is rarely admitted in public, the vast majority of bank regulatory authorities of industrialized countries follow "too-big-to-fail" (TBTF) policies (Roth (1994)). The negative macroeconomic consequences generated by the failure of a large financial institution make TBTF key issue to be considered in any country. Critics of TBTF argue that the doctrine is unfair to smaller banks because it provides an incentive for larger banks to increase risk, for example, by maintaining a lower capital-to-asset ratio than smaller banks. There is another point about TBTF policies when applied to transition economies: the largest banks are still at least partly in state hands in many countries. Therefore, the size of the bank is also a proxy for its ownership structure.

In the present paper we study models of bank failure in the context of transition economies. Data from the Czech banking crisis (1994 to1996) will be used to test our approach empirically. We expect that only a small group of variables used to predict bank failure in mature markets will actually help explain bank failures in transition economies. On the other hand, we expect that the quality of auditor used ("Big Six" versus local firms) will provide valuable additional information. It should be in fact, an indicator of whether we can use the balance-sheet data from the conventional models at all. Our central hypothesis is that the retail deposit interest rate in transition economies could be used as a proxy to reflect the default risk of a bank, and therefore this information should improve the quality of bank failure prediction .

The paper is organized as follows: section 2 introduces additional variables that should be used for predicting bank failure in transition economies. Section 3 describes the emergence of the Czech banking sector; section 4 present the results; and the final section containes conclusions and policy prescriptions.

4.2 Improving "Standard" Models of Bank Failures

The vast majority of bank failure models group variables according to **CAMEL**, an acronym for Capital adequacy, Asset quality, Management, Earnings and Liquidity, and/or use CAMEL rankings. In selecting variables that might influence bank failure in transition economies, we suggest also considering regulatory factors, that is, those describing the requirements of supervisory bodies such as minimum reserve requirements, as well as economic factors. Such economic factors can significantly interfere with CAMEL factors. Indeed, an economic upturn could be expected to increase the likelihood of bank survival and bank survival time, primarily through associated increases in asset quality. In our analysis we did not analyze the relationship between bank failure and economic factors, since during the period studied the Czech Republic was experiencing constant growth, and, therefore, the worsening of banks' loan portfolios had nothing to do with worsening economic conditions. Usually local banks in transition economies are exposed to very high risk. This risk results from lending money to borrowers with a short or nonexistent track record in business whose business plans must be realized in a rapidly changing market environment. Moreover, most bank staffs responsible for credit rationing are young, inexperienced, and potentially corrupt. It is generally expected that for troubled banks large certificate of deposits (portions of which are not explicitly insured) are a less stable and potentially more expensive funding source than retail deposits. In particular, low liquidity is often associated with aggressive strategies and high-risk profiles. Marcus (1990) predicts a tendency for individual banks to converge towards either a high-risk or low-risk posture, depending on their charter value. We expect similar behavior in transition economies; smaller banks would more likely adopt aggressive market strategies and/or a high-risk posture.

Ellis and Flannery (1992) analyze how the spread between large certificates of deposit (CD) and Treasury bill rates relates to the banks default risk. Like other studies using CD rates, they conclude that measured bank-risk influenced CD rates. It is difficult to make a direct link between CD rates and the default risk, however, because the spread also reflects other factors such as market imperfection, differences in liquidity and TBTF. Nevertheless, we expect that a link exists because the benefits of market-based regulatory policies would be quite limited if large banks' default risks were not priced.

Unfortunately, we cannot directly use this approach because the CD market is generally missing in transition economies, so we suggest using standard retail deposit rates instead of CD rates. We expect that retail deposit rates should linked in a similar way to default risk as CD rades (See Figure 1 and 2). Key issues for bank regulators were the quality of independent auditors and the accounting standards employed. One should keep in mind that local accounting standards in transition economies have roots in central planning, and therefore are tailored to report production rather than profit.[1] There exist several problems associated with Czech accounting standards that seriously affect a bank's financial positioning. One of the most dangerous is the lack of charge-offs for loan losses.[2] It is striking that banks cannot charge-off loan losses before a borrower files for bankruptcy. Not surprisingly, then, a bank could report a profit even if it in fact experiences a loss.

It is clear that the quality and independence of external auditors is highly correlated with both the effectiveness of regulators and the quality of early warning signals.[3] Unfortunately, for the purpose of our analysis we cannot include proxies for auditor quality, since certain auditors would *ex post* form a perfect proxy for predicting bank failure in the Czech Republic. It is striking that prior to 1994, the Czech National Bank (CNB) did not require banks to use only pre-selected auditors (say the Big Six/Five).[4]

We expect that adding the retail interest rate variable into a model of bank failure should significantly help with the balance-sheet-data problems mentioned above. Basically, even when balance sheets show very flattering figures and results, insiders (i.e., other banks) know which banks are more likely to fail. A bank in this position has limited access to loans from other banks. In order to maintain liquidity, the bank needs to attract cash via a much higher interest rate on term deposits than other (i.e., "safe") banks offer. Unfortunately, this only speeds up the process of worsening bank conditions. Cordella and Yeyati (1998) show that when banks do not control their risk exposure, the presence of informed depositors may increase the probability of bank failures. In the case of the Czech Republic a clear pattern for one-year deposits is seen (see Figure 1), indicating significant liquidity problems of problematic

---

[1] It is striking how much local and international accounting standards differ. For instance, in 1992 Komercni banka reported a profit of 3.2 billion CZK and a loss of 5.9 billion CZK according to local and international standards, respectively.

[2] Probably the original reason was to avoid a significant tax reduction. Similarly, tax treatment of losses and reserve creation was changed several times.

[3] One has to keep in mind that the vast majority of early warning signals are based on balance sheet data.

[4] Outside observers heavily criticized the regulators: "A litany of scandal related to investment funds and more forced supervision by the CNB made the country's financial sector look like it was run by Mickey Mouse with the rest of the Disney crew doing the audits" Czech Business Journal, (May/June 1996).

banks and either lack of capital or credibility that would allow them to get financing via the inter-bank market or other refinancing instruments.

4.3   Czech banking sector

The first step in reforming the banking sector was the law creating a central bank, the State Bank of Czechoslovakia (hereafter SBCS), No. 130/1989, approved on November 15, 1989. According to this law, the SBCS was responsible for state monetary policy, but not for commercial banking. The law that regulates the commercial banks and savings-and-loans sector was approved one month later (December 13, 1989). This law enabled two-tier banking, in that it brought into being commercial banks and set the basic rules for their operation. Banking sector regulation was exercised by the Ministry of Finance.[5] According to this law, interest rates were governed by the SBCS and deposits in state financial institutions were guaranteed by the state. In January 1990, the SBCS transferred its commercial banking to three newly established banks: Komerčni banka (KB), Vseobecka uverova (VUB) and Investicni banka (IP, which in 1993 merged with Post office banks as IPB). On December 20, 1991 new laws on central and other banks were adopted (Nos. 21 and 22/1992). These laws, effective from February 1, 1993, established the independence of the national bank from the government and gave the SBCS the authority for banking supervision. On January 1, 1993, the Czech National Bank (CNB) took over the functions of the SBCS as a result of the split of Czechoslovakia. The law on banks also had clearly specified rules for granting licenses, and set up a general regulatory framework .

Unfortunately, the initial conditions for obtaining banking licenses were quite soft, requiring a minimum subscribed equity capital of only CZK 50 million (US$2 million). This low requirement was increased in April 1991 to CZK 300 million (US$10 million). On the other hand, the local market was protected against foreign competition by the "Law on Foreign Exchange," which prevented firms from directly acquiring capital abroad. With such a low capital requirements the number of new banks literally exploded in early 1990. While in early 1990 there was a central bank plus seven banks licensed for universal banking, by the end of 1990 there were 23. This trend continued with 36 banks by the end of 1991 and 51 by the end of 1992. These newly established banks were, in general, small, with Agrobanka being the one significant

---

[5]The Federal Ministry of Finance supervised banks and the Ministries of Finance of the Czech and Slovak Republics controlled savings and loans.

exception.[6] In 1993, the rate of new bank creation slowed, with only 8 new banking licenses granted. Between mid-1994 and 1996, the CNB decided not to grant any new bank licenses, most likely in response to failures of small and medium banks. The CNB probably expected that a lack of new bank licenses would cause the banking sector to consolidate through mergers/acquisitions of smaller and troubled banks. This policy decision was only partially successful, and the CNB started to grant new bank licenses again in 1996. Neither big Czech banks nor Czech branches of foreign banks were enthusiastic about buying their bankrupted competitors. Similarly, for foreign banks not yet present on the Czech market, such an acquisition would be equivalent to a new bank license, and since the asset quality of troubled banks was bad, the resulting price for a "new license" was very high.

Due to very soft licensing procedures and insufficient screening of license candidates, many newly-formed banks lacked a sufficient capital base, as well as employees equipped with proper managerial skills and business ethics. Because of their lack of capital, all small and medium-sized banks had to deal with an adverse selection problem. With their lending rates the highest in the market, the overhelming majority of their clients became those undertaking the riskiest projects, those which other banks refused to finance. In addition, several new banks were using deposits to extend credit to other activities of the bank's owners, or simply "tunneling" the deposited money out of the bank. Regardless of whether the main reason was incompetence or theft, the overall effect on the cash flow and balance sheets of these banks was damaging.[7] Beginning in December 1993 several bank failures disturbed public trust in the banking sector and had a strong influence on the stability of small and medium-sized banks.

As a reaction to the first three bank failures, the Law on Banks was amended to include obligatory insurance on deposits. This insurance covered only deposits of citizens up to 100,000 CZK per head and per bank, with the premium being limited to 80 percent of the deposit balance on the day of a bank's closure. The amendment also increased the extent and authority of banking supervision granted to the CNB. The CNB could now impose sanctions for noncompliance with its regulations ranging from enforcing corrections and imposing fines to the revocation of banking licenses.[8]

---

[6] Agrobanka, founded in 1990, became the fifth largest bank in the Czech Republic within a year.

[7] The Economist (September 1996): "Each of these bank failures stemmed from a deadly cocktail of mismanagement, orgiastic lending (often to a bank's own stockholders), and more often than not, fraud."

[8] The CNB has been given the authority to 1) force banks to fulfill several obligatory

After the introduction of deposit insurance, another bank, Česka banka, filed for bankruptcy and the new law was applied to its clients. However, when a series of additional failures soon followed in the 1996 election year, the CNB became far more generous, with individual clients of the failed banks recovering all of their deposits up to 4,000,000 CZK, in contradiction to the law's provisions. The CNB decided to cope with the resulting sensitive political problem of lost deposits by tightening the licensing procedures and modifying obligatory deposit insurance. In its efforts to stem the tide of bank failures, the CNB tried two policies. In early July 1995, it tightened its policies, increasing the minimum reserve requirements (MRR) and also unifying its rates. Until then, checking account deposits had a MRR rate three times that of term deposits. The level and rapid changes of MRR suggest that this instrument had been used in the past quite heavily as a monetary tool. It is clear that such a high level of the MRR coupled with the feature that these reserves earn no interest had a significant effect on the direct regulation of money in circulation.[9] Although such a setting was useful from the standpoint of regulating monetary development, it created a substantial financial burden for commercial banks. Banks with a higher proportion of term deposits (namely the Czech Savings Bank and some smaller financial institutions) were particularly hurt by this measure. One can speculate and make connections between changes of MRR over time and actual bank failures.[10]

4.4 Data and the Results

The major obstacle to this project was to get reliable data. For financial indicators the only set of publicly available accounting data is represented by a subset of the ASPEKT (or CEKIA) databases of the Czech capital market that covers annual reports of publicly traded banks. Unfortunately, these data sets are basically useless for applying standard models of bank failure for two reasons. First, the publicly available information covers only a short version of the balance sheet. Second, if any additional public information exists (for example, a standard bal-

---

rules, 2) approve/change bank management, 3) impose a penalty up to 50 mil. CZK , 4) enforce reduction of shareholder's capital and its transfer to reserves if these were not sufficient , and 5) withdraw or freeze banking licenses.

[9] During the period 1990 to 1992 MRR carried 4% interest.

[10] Clearly an increase of zero-interest MRR would affect both the probability of bank failure and survival time. Let us mention that an MRR rate of 9.5% and zero interest resulted in an approximate 1.6 percentage point change on the interest rate margin. Therefore, domestic bank competitiveness was significantly influenced since foreign banks carry lower MRR.

ance sheet provided by ASPEKT or CEKIA), then several variables are missing, namely for those banks that were *ex-post* seen as "problematic."

In order to minimize these data problems we use financial variables/ratios that were constructed, instead of the original data held by the CNB regulatory body. Although we construct the financial ratios that have been used in similar studies, we must stress that our indicators do not have the same meaning as in the other studies, since all reporting to the CNB was done according to local accounting standards. Our data set covers 20 local banks of which 14 posed significant problems at some point during the study. It does not make any sense to add foreign banks or their branches to our sample because of their inherent differences in terms of services, structure, financing, etc. As discussed earlier, we expect to see a gap in retail deposit rates between sound and problematic banks. As shown in Figures 1 and 2, banks with higher interest rates on term deposits were more likely to fail. This finding is in line with our original objective of capturing the default risk using retail deposit rate as a proxy. Although Figure 1 indicates a strong pattern, we want to test whether the differences between groups are significant. Table 1 summarizes several t-tests across different time periods and maturities. These tests empirically verify that (mean) interest rates for problematic banks were significantly higher compared with sound banks. In addition, we see that since the first half of 1994, differences were statistically significant for both one and two year term deposits.[11] Moreover, the mean difference was higher for longer maturities, a finding that is consistent with the idea of capturing default risk for the bank via retail deposit rates.

The next step in our analysis was to compare quality of bank failure prediction with and without information on the retail deposit interest rates. Results of logit models are presented in Table 2. Note that selection of other variables or probit specification gave similar results. First, the financial indicators, although they were drawn from official data collected by the supervisory body of the CNB (used in Model 1), did not provide significantly better prediction of actual bank failure than one-year deposit rates alone (Model 2).[12] This finding suggests that (non-audited) balance sheets with detailed information used by the supervisory body

---

[11] We suggest excluding checking accounts from our analysis. Usually, the interest rate on those accounts is not a relevant measure of why clients opted for the particular bank. (We omit whole range of services offered by the bank.).

[12] Previous t-tests suggest to using two-year deposit rates, although the reason why we used one-year rates instead is simple. For two-year deposit rates we have a few missing observations: not every bank provided a table of retail interest rates by all maturities and several banks specified longer maturities as "negotiable". Since we do not want to lose more observations, we opted for one-year interest rate that was provided by all banks in our sample.

of the CNB did not contain more information with respect to prediction of actual bank failure than publicly available interest rate data. More importantly, there is an interesting interaction between information contained in financial ratios and retail interest rates. Looking at the results for the first half of 1995 we see that, although both Models 1 and 2 provide very similar (and not very good) predictions, combining them (Model 3) significantly increases the quality of our predictions. In addition, when analyzing the second half of 1995, we see an interesting change: a significant difference no longer exists in the quality of prediction between models 1 and 3. In other words, this result means that information in retail deposit rates is already reflected in the bank balance sheet. One can then speculate and make a connection between an actual upcoming bank failure and seemingly indefensible financial situation of the bank, revealed via "the sudden appearence of" significant red numbers on balance sheets.

4.5 Conclusions

Despite the small sample available, we would like to highlight a few lessons that can be learned from the Czech banking crisis. Our study stresses the significant role played by auditors and local accounting standards. The magnitude of shock adjustments of the minimum reserve requirements opens up the question of to what extent actual bank failures were affected by policies of the supervisory body of the CNB. Because of the same timing of changes that affected the accounting of reserve creation, charges against losses, etc. and adjustments of MRR, we cannot distinguish among those effects.[13] Our results also show that, in the early years of transition, the supervisory body did not have much better information with respect to predicting bank failure than that which was available to the general public via retail interest rates. Finally, our results suggest that it would be useful to combine balance sheet and interest rate data. In the Czech case this significantly improved the quality of the prediction of bank failures.

---

[13]The EBRD (1997)transition report mentioned that tightening of central bank credit precipitated the liquidity crisis in several transition countries. As a result, the number of banks in Russia declined from a peak of 2,561 in 1995 to 1,887 by June 1997. In the Baltic states the number of banks fell in the aftermath of banking crises from a peak of 43 (1992) to 14 (1996) in Estonia, 61 (1993) to 33 (1996) in Latvia, and 27 (1994) to 12 (1996) in Lithuania.

Table 1. Comparison of average deposit rates (control group vs. problematic banks). Semiannual data from June 1993 to December 1995

Year	Group	Checking account	1 year term deposit	2 year term deposit
1993.1	"control"	3.48	12.95	14.53
	"problematic"	4.44	13.74	14.36
p-value (t-tests)		.05**	0.16	0.4
1993.2	"control"	3.97	13.05	14.44
	"problematic"	4.29	13.95	14.72
p-value (t-tests)		0.27	.03**	0.22
1994.1	"control"	3.61	10.51	13.42
	"problematic"	4.21	11.75	13.6
p-value (t-tests)	0.18	.05**	.00***	
1994.2	"control"	3.4	9.82	12.83
	"problematic"	4.36	10.80	14.51
p-value (t-tests)		.08*	.10*	.05**
1995.1	"control"	3.17	9.45	11.68
	"problematic"	4.47	10.61	13.67
p-value (t-tests)		.01***	.03**	.00***
1995.2	"control"	3.56	9.62	11.15
	"problematic"	4.68	10.63	12.92
p-value (t-tests)		.03**	.01***	.01***

*** Significant at 1% level, ** Significant at 5% level,
* Significant at 10% level

Table 2. Comparison of logit models. Standard errors are in parentheses.

Variable	Period 1995/1			Period 1995/2		
	I.	II.	III.	I.	II.	III.
CA, Capital adequacy	.10 (.65)		-.53 (.39)	.17 (.11)		-.54 (.49)
EM, Equity multiplier	-.05 (.09)		2.57 (1.6)	-.04 (.11)		-.1.4 (1.2)
ROA, Return on Assets	-.31 (.22)		-.82 (.61)	-2.3 (.56)		-.01 (.88)
LLRCL, Classified Loans Coverage by Provisions	.05 (.07)		-.83 (.70)	-1.9 (1.3)		-3.0 (2.9)
Y1_H, One Year Term deposit rate (the highest)		.15 (.11)	2.68 (1.8)		.07 (.14)	1.3 (1.5)
Y1_L, One Year Term deposit rate (the lowest)		-.14 (.18)	-.85 (.84)		-.02 (.22)	.49 (1.1)
R-square	0.1	0.09	0.71	0.35	0.06	0.69
Fraction of Correct Prediction	0.65	0.7	0.9	0.75	0.63	0.88
Test I. vs. II. (p-value)	$\chi^2(1) = 0.20$ (.65)			$\chi^2(1) = 0.67$ (.41)		
Test II. vs. III. (p-value)	$\chi^2(1) = 2.67$ (.10)*			$\chi^2(1) = 2.67$ (.10)*		
Test I. vs. III. (p-value)	$\chi^2(1) = 3.57$ (.06)*			$\chi^2(1) = 1.0$ (.32)		

** Significant at 5% level, * Significant at 10% level.

$^+$The test reported here is a chi-square test whether one model dominates the other in terms of accuracy of prediction. Our null hypothesis is that there is no difference between those models in prediction accuracy. Denote by $"+"$ the cases when models correctly predicted the dependent variable, and by $"--"$ when they did not. The quality of the prediction can then be summarized in the following table:

	Model 1	+	−	Σ
Model 2	+	n_{11}	n_{12}	$n_{1.}$
	−	n_{21}	n_{22}	$n_{2.}$
	Σ	$n_{.1}$	$n_{.2}$	n

Corresponding test statistic

$$\chi^2 = \frac{(n_{12} - n_{21})^2}{n_{12} + n_{21}}$$ has chi-square distribution with 1 degree of freedom.

For more details see Hanousek (1998).

Figures 1 and 2. One Year Retail Deposit Rates (the highest)

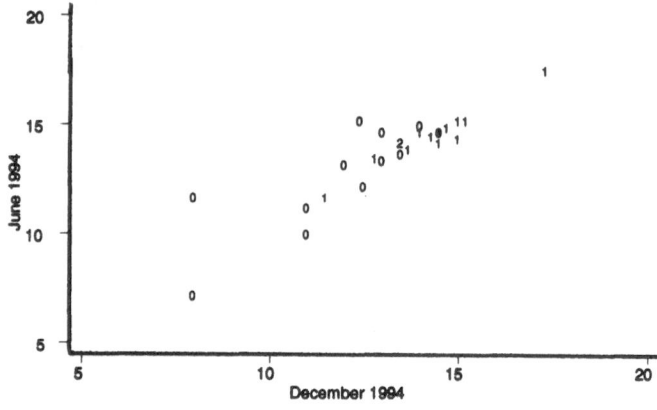

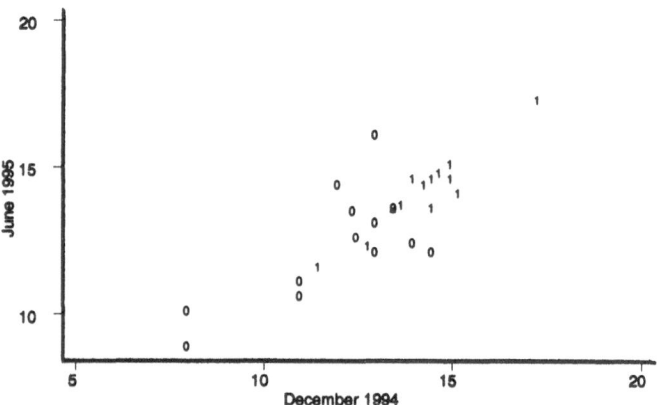

Symbol "1" indicates the banks that failed during the period 1994-1996, while "0" denotes those banks which "survived". As a benchmark (denoted by 2) we used "Plzenska banka", the bank that provided practically no corporate lending, and therefore, their interest rates should not reflect problems with their portfolio of loans.

Bibliography

Barber, J. R., Chang, C. and Thurston, D. (1996). Bank failure, risk, and capital regulation, *Journal of Economics and Finance* **3**(20): 13–20.

Cordella, T. and Yeyati, E. (1998). Public disclosure and bank failures, *IMF Staff Papers* **1**(45).

Ellis, D. and Flannery, M. (1992). Does the debt market assess large banks, *Journal of Monetary Economics* (30): 481–502.

Hanousek, J. (1998). Specification tests of (binary) choice models. a non-parametric approach, *Proceedings of Prague Stochastics*, pp. 213−−216.

Hwang, D., Lee, C. and Liaw, K. (1997). Forecasting bank failures and deposit insurance premium, *International Review of Economics and Finance* **3**(6): 317−−34.

Lane, W., Looney, S. and Wansley, J. (1990). An application of the Cox proportional hazard model to bank failure, *Journal of Banking and Finance* pp. 511–531.

Looney, S., Wansley, J. and Lane, W. (1989). An examination of mis-classification with bank failure prediction models, *Journal of Economics and Business* .

Marcus, A. J. (1990). Deregulation and bank financial policy, *Journal of Banking and Finance* (8): 557–565.

Roth, M. (1994). Too-big-to-fail and the stability of the banking system: Some insights from foreign, *Business Economics Countries* pp. 43–49.

5 Credit Scoring using Semiparametric Methods

Marlene Müller and Bernd Rönz

5.1 Introduction

Credit scoring methods aim to assess credit worthiness of potential borrowers to keep the risk of credit loss low and to minimize the costs of failure over risk groups. Typical methods which are used for the statistical classification of credit applicants are linear or quadratic discriminant analysis and logistic discriminant analysis. These methods are based on scores which depend on the explanatory variables in a predefined form (usually linear). Recent methods that allow a more flexible modeling are neural networks and classification trees (see e.g. Arminger, Enache and Bonne, 1997) as well as nonparametric approaches (see e.g. Henley and Hand, 1996).

Logistic discrimination analysis assumes that the probability for a "bad" loan (default) is given by $P(Y = 1|X) = F(\beta^T X)$, with $Y \in \{0,1\}$ indicating the status of the loan and X denoting the vector of explanatory variables for the credit applicant. We consider a semiparametric approach here, that generalizes the linear argument in the probability $P(Y = 1|X)$ to a partial linear argument. This model is a special case of the Generalized Partial Linear Model $E(Y|X,T) = G\{\beta^T X + m(T)\}$ (GPLM) which allows to model the influence of a part T of the explanatory variables in a nonparametric way. Here, $G(\bullet)$ is a known function, β is an unknown parameter vector, and $m(\bullet)$ is an unknown function. The parametric component β and the nonparametric function $m(\bullet)$ can be estimated by the quasilikelihood method proposed in Severini and Staniswalis (1994).

We apply the GPLM estimator mainly as an exploratory tool in a practical credit scoring situation. Credit scoring data usually provide various discrete and continuous explanatory variables which makes the application of a GPLM interesting here. In contrast to more general nonparametric approaches, the estimated GPLM models allow an easy visualization and interpretation of the results. The estimated curves indicate in which direction the logistic discriminant should be improved to obtain a better separation of "good" and "bad" loans.

The following Section 5.2 gives a short data description. Section 5.3 presents the results of a logistic discrimination analysis. Section 5.4 describes the semiparametric extension to the logistic discrimination analysis. We estimated and compared different variations of the semiparametric model in order to see how the several explanatory variables influence credit worthiness. Section 5.5 compares the semiparametric fits the classic logistic analysis. Finally, Section 5.6 discusses the estimated models with respect to misclassification and performance curves.

5.2   Data Description

The analyzed data in this paper have been provided by a French bank. The given full estimation sample (denoted as **data set A** in the following) consists of 6672 cases (loans) and 24 variables:

- Response variable Y (status of loan, 0="good", 1="bad"). The number of "bad" loans is relatively small (400 "bad" versus 6272 "good" loans in the estimation sample).

- Metric explanatory variables X2 to X9. All of them have (right) skewed distributions. Variables X6 to X9 in particular have one realization which covers a majority of observations.

- Categorical explanatory variables X10 to X24. Six of them are dichotomous. The others have three to eleven categories which are not ordered. Hence, these variables need to be categorized into dummies for the estimation and validation.

Figure 5.1 shows kernel density estimates (using rule-of-thumb bandwidths) of the metric explanatory variables X2 to X9. All density estimates show the existence of outliers, in particular in the upper tails. For this reason we restricted our analysis to only those observations with X2, ... , X9 $\in [-3, 3]$. We denote the resulting data set of 6180 cases as **data set B**. The kernel density estimates for this smaller sample are shown in Figure 5.2. Figure 5.3 shows some bivariate scatterplots of the metric variables X2 to X9. It can be clearly seen that the variables X6 to X9 are of quasi-discrete structure. We will therefore concentrate on variables X2 to X5 for the nonparametric part of semiparametric model.

In addition to the estimation sample, the bank provided us with a validation data set of 2158 cases. We denote this validation data set as **data set C** in the following. Table 5.1 summarizes the percentage of "good" and "bad" loans in each subsample.

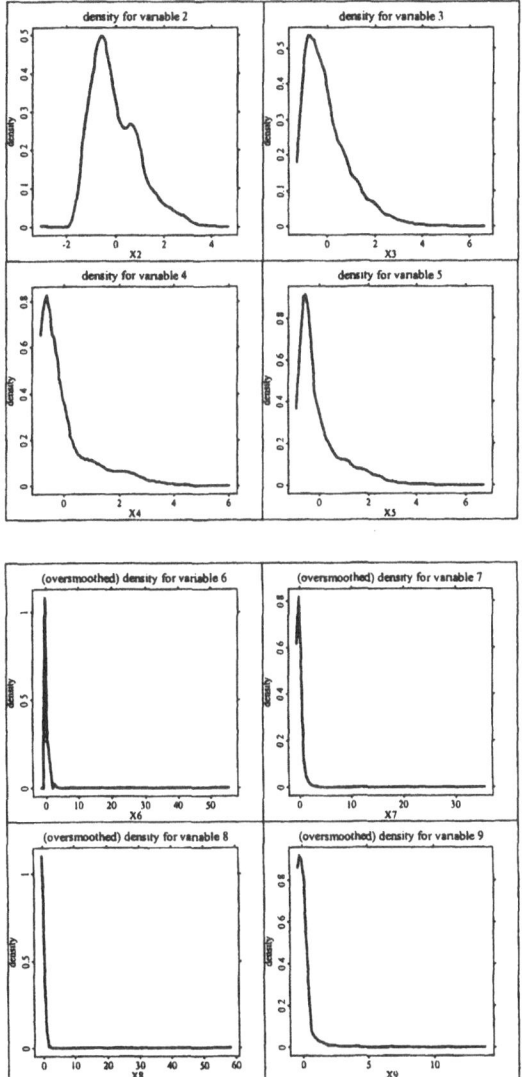

Figure 5.1: Kernel density estimates, variables X2 to X9, estimation
data set A.

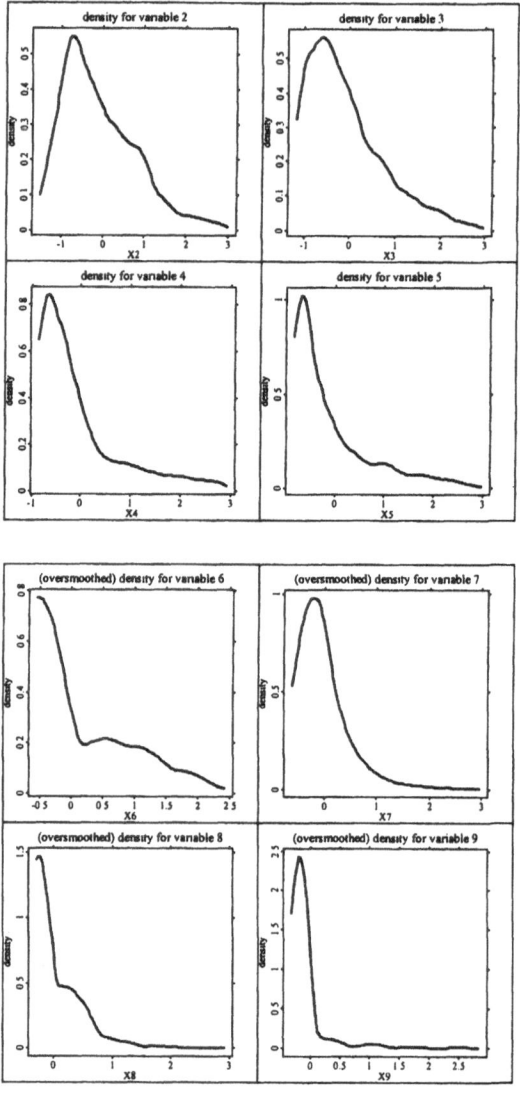

Figure 5.2: Kernel density estimates, variables X2 to X9, estimation data set B.

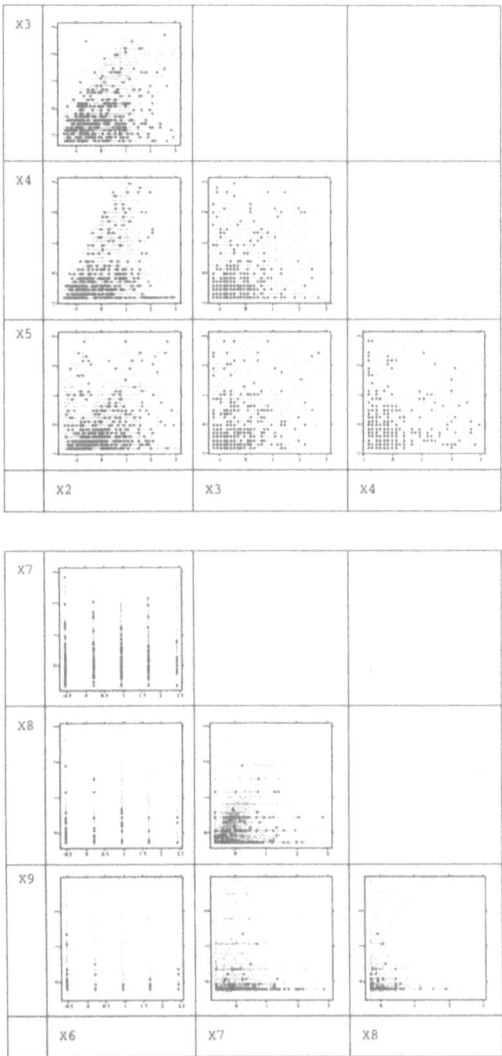

Figure 5.3: Scatterplots, variables X2 to X5 (upper plot) and X6 to X9 (lower plot), estimation data set B. Observations corresponding to Y=1 are emphasized in black.

5.3 Logistic Credit Scoring

The logit model (logistic discriminant analysis) assumes that the probability for a "bad" loan is given by

$$P(Y = 1|X) = F\left(\sum_{j=2}^{24} \beta_j^T X_j + \beta_0\right) \qquad (5.1)$$

	Estimation (full) data set A		Estimation (used) data set B		Validation data set C	
0 ("good")	6272	(94.0%)	5808	(94.0%)	2045	(94.8%)
1 ("bad")	400	(6.0%)	372	(6.0%)	113	(5.2%)
total	6672		6180		2158	

Table 5.1: Responses in data sets A, B and C.

where

$$F(u) = \frac{1}{1 + \exp(-u)}$$

is the logistic (cumulative) distribution function. X_j denotes the j-th variable if Xj is metric ($j \in \{2, \dots, 9\}$) and the vector of dummies if Xj is categorical ($j \in \{10, \dots, 24\}$). For all categorical variables we used the first category as reference.

The logit model is estimated by maximum–likelihood. Table 5.2 shows the estimation results for this model. It turns out, that in fact all variables contribute more or less to the explanation of the response. The modeling for the categorical variables cannot be further improved, since by using dummies one considers all possible effects. Concerning the continuous variables, we observe nonsignificant coefficients for some regressors. The continuous variables get more attention by using semiparametric models.

5.4　Semiparametric Credit Scoring

The logit model (5.1) is a special case of the generalized linear model (GLM, see McCullagh and Nelder, 1989) which is given by

$$E(Y|X) = G(\beta^T X).$$

In the special case of a binary response we have

$$E(Y|X) = P(Y = 1|X).$$

The semiparametric logit model that we consider here generalizes the linear argument $\beta^T X$ to a partial linear argument:

$$E(Y|X, T) = G\{\beta^T X + m(T)\}$$

This generalized partial linear model (GPLM) allows us to describe the influence of a part T of the explanatory variables in a nonparametric way.

Variable	Coefficient	S.E.	t-value	Variable	Coefficient	S.E.	t-value
X0 (const.)	**-2.605280**	0.5890	-4.42	X19#2	-0.086954	0.3082	-0.28
X2	**0.246641**	0.1047	2.35	X19#3	0.272517	0.2506	1.09
X3	**-0.417068**	0.0817	-5.10	X19#4	-0.253440	0.4244	-0.60
X4	-0.062019	0.0849	-0.73	X19#5	0.178965	0.3461	0.52
X5	-0.038428	0.0816	-0.47	X19#6	-0.174914	0.3619	-0.48
X6	**0.187872**	0.0907	2.07	X19#7	0.462114	0.3419	1.35
X7	-0.137850	0.1567	-0.88	X19#8	**-1.674337**	0.6378	-2.63
X8	**-0.789690**	0.1800	-4.39	X19#9	0.259195	0.4478	0.58
X9	**-1.214998**	0.3977	-3.06	X19#10	-0.051598	0.2812	-0.18
X10#2	-0.259297	0.1402	-1.85	X20#2	-0.224498	0.3093	-0.73
X11#2	**-0.811723**	0.1277	-6.36	X20#3	-0.147150	0.2269	-0.65
X12#2	-0.272002	0.1606	-1.69	X20#4	0.049020	0.1481	0.33
X13#2	0.239844	0.1332	1.80	X21#2	0.132399	0.3518	0.38
X14#2	-0.336682	0.2334	-1.44	X21#3	**0.397020**	0.1879	2.11
X15#2	**0.389509**	0.1935	2.01	X22#2	-0.338244	0.3170	-1.07
X15#3	0.332026	0.2362	1.41	X22#3	-0.211537	0.2760	-0.77
X15#4	**0.721355**	0.2580	2.80	X22#4	-0.026275	0.3479	-0.08
X15#5	0.492159	0.3305	1.49	X22#5	-0.230338	0.3462	-0.67
X15#6	**0.785610**	0.2258	3.48	X22#6	-0.244894	0.4859	-0.50
X16#2	**0.494780**	0.2480	2.00	X22#7	-0.021972	0.2959	-0.07
X16#3	-0.004237	0.2463	-0.02	X22#8	-0.009831	0.2802	-0.04
X16#4	0.315296	0.3006	1.05	X22#9	0.380940	0.2497	1.53
X16#5	-0.017512	0.2461	-0.07	X22#10	-1.699287	1.0450	-1.63
X16#6	0.198915	0.2575	0.77	X22#11	0.075720	0.2767	0.27
X17#2	-0.144418	0.2125	-0.68	X23#2	-0.000030	0.1727	-0.00
X17#3	**-1.070450**	0.2684	-3.99	X23#3	-0.255106	0.1989	-1.28
X17#4	-0.393934	0.2358	-1.67	X24#2	0.390693	0.2527	1.55
X17#5	**0.921013**	0.3223	2.86				
X17#6	**-1.027829**	0.1424	-7.22				
X18#2	0.165786	0.2715	0.61				
X18#3	0.415539	0.2193	1.89				
X18#4	**0.788624**	0.2145	3.68				
X18#5	**0.565867**	0.1944	2.91	df			6118
X18#6	0.463575	0.2399	1.93	Log-Lik.			-1199.6278
X18#7	**0.568302**	0.2579	2.20	Deviance			2399.2556

Table 5.2: Results of the Logit Estimation. Estimation data set B. Bold coefficients are significant at 5%.

Here, $G(\bullet)$ is a known function, β is an unknown parameter vector, and $m(\bullet)$ is an unknown function. The parametric component β and the nonparametric function $m(\bullet)$ can be estimated by the quasilikelihood method proposed in Severini and Staniswalis (1994).

We will use the GPLM estimator mainly as an exploratory tool in our practical credit scoring situation. Therefore we consider the GPLM for several of the metric variables separately as well as for combinations of them. As said before, we only consider variables X2 to X5 to be used within a nonparametric function because of the quasi–discrete structure of X6 to X9. For instance, when we include variable X5 in a nonlinear way, the parametric logit model is modified to

$$P(Y = 1|X) = F\left(m_5(X_5) + \sum_{j=2, j \neq 5}^{24} \beta_j^T X_j\right)$$

where a possible intercept is contained in the function $m_5(\bullet)$.

Table 5.3 contains only the parametric coefficients for the parametric and semiparametric estimates for variables X2 to X9. The column headed by "Logit" repeats the parametric logit estimates for the for model with variables X2 to X24. The rest of the columns correspond to the semi-parametric estimates where we fitted those variables nonparametrically which are heading the columns.

| Variable | Logit | Nonparametric in | | | | | |
		X2	X3	X4	X5	X4,X5	X2,X4,X5
constant	-2.605	–	–	–	–	–	–
X2	0.247	–	0.243	0.241	0.243	0.228	–
X3	-0.417	-0.414	–	-0.414	-0.416	-0.408	-0.399
X4	-0.062	-0.052	-0.063	–	-0.065	–	–
X5	-0.038	-0.051	-0.045	-0.034	–	–	–
X6	0.188	0.223	0.193	0.190	0.177	0.176	0.188
X7	-0.138	-0.138	-0.142	-0.131	-0.146	-0.135	-0.128
X8	-0.790	-0.777	-0.800	-0.786	-0.796	-0.792	-0.796
X9	-1.215	-1.228	-1.213	-1.222	-1.216	-1.214	-1.215

Table 5.3: Parametric coefficients in parametric and semiparametric logit, variables X2 to X9. Estimation data set B. Bold values are significant at 5%.

It turns out, that all coefficients vary little over the different estimates. This holds as well for their significance (determined by a t–test). Variables X4 and X5 are constantly insignificant over all estimates. Hence, they are interesting candidates for a nonparametric modeling: variables which are significant may already capture a lot of information on Y by the parametric inclusion into the model.

The semiparametric logit model is estimated by semiparametric maximum-likelihood, a combination of ordinary and smoothed maximum-li-kelihood. The fitted curves for the nonparametric components according

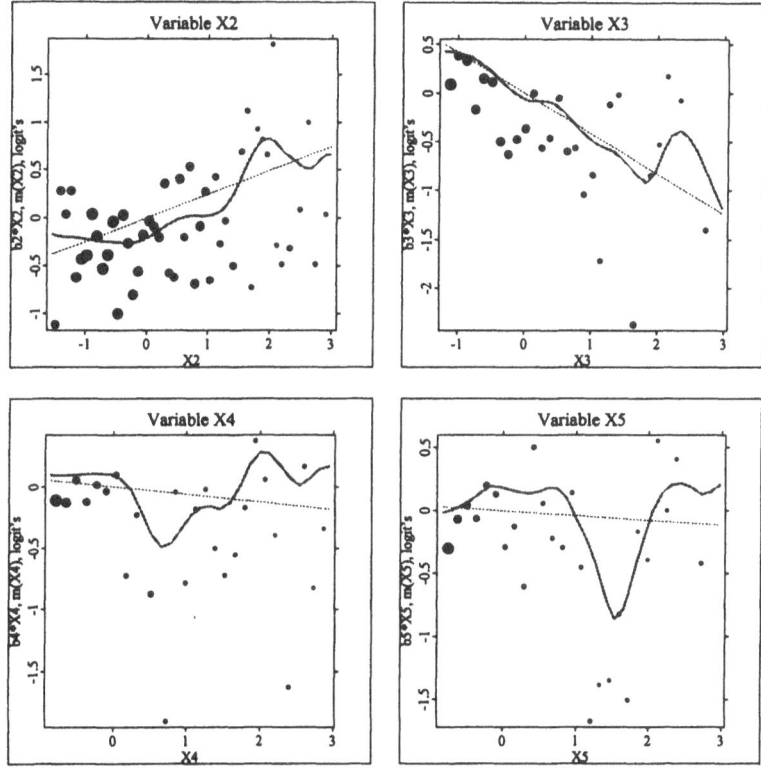

Figure 5.4: Marginal dependencies, variables X2 to X5. Thicker bullets correspond to more observations in a class. Parametric logit fits (thin dashed linear functions) and GPLM logit fits (thick solid curves).

to Table 5.3 can be found in Figures 5.4 for the marginal fits (variables X2 to X5 separately as the nonparametrical component) and Figure 5.6 for the bivariate surface (variables X4 and X5 jointly nonparametrically included). Additionally, Figures 5.4 and 5.5 reflect the actual dependence of the response Y on variables X2 to X9. We have plotted each variable restricted to [-3,3] (i.e. the data from sample B) versus the logits

$$logit = \log \left(\frac{\widehat{p}}{1 - \widehat{p}} \right)$$

where $\widehat{p}$ are the relative frequencies for $Y = 1$. Essentially, these logits are obtained from classes of identical realizations. In case that $\widehat{p}$ was 0 or 1, several realizations have been summarized into one class. For all variables but X7 this only concerns single values.

The plots of the marginal dependencies for variables X6 to X9 show that the realizations essentially concentrate in one value. Hence we did not fit a nonparametric function here.

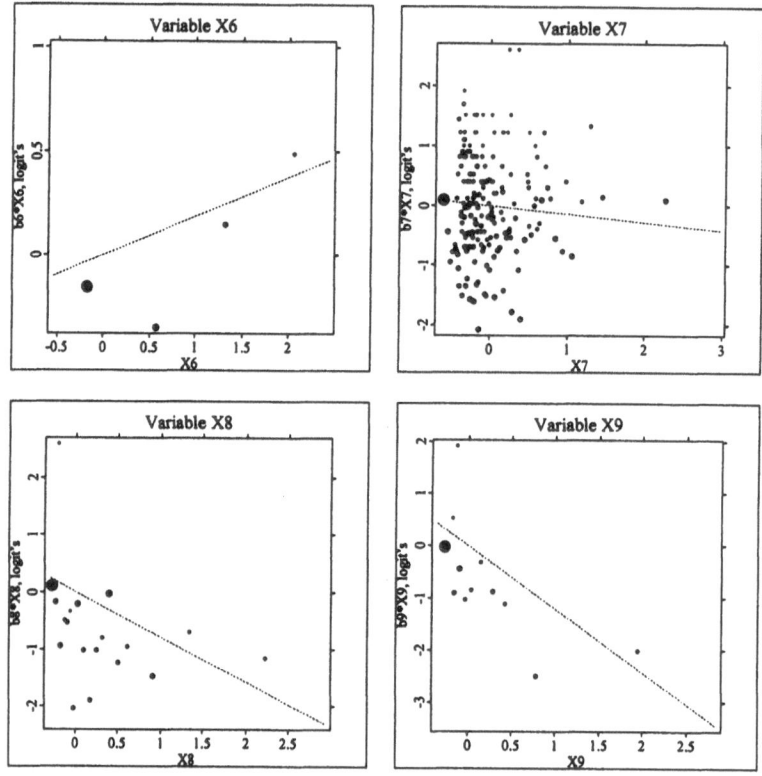

Figure 5.5: Marginal dependencies, variables X6 to X9. Thicker bullets correspond to more observations in a class. Parametric logit fits (thin dashed). Estimation data set B.

5.5   Testing the Semiparametric Model

To assess, whether the semiparametric fit outperforms the parametric logit or not, we have a number of statistical characteristics. For the above estimated models, they are summarized in Table 5.4.

The deviance is minus twice the estimated log–likelihood of the fitted

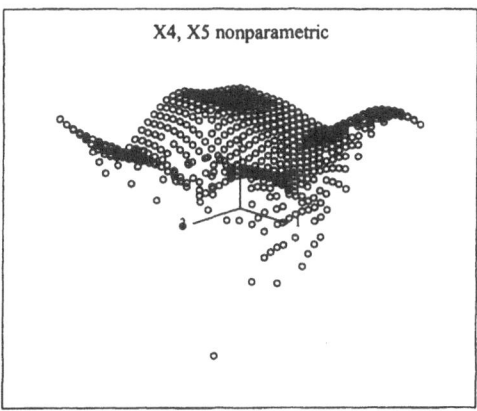

Figure 5.6: Bivariate nonparametric surface for variables X4, X5. Estimation data set B.

model in our case. For the parametric case, the degrees of freedom just denote

$$df = n - k$$

where n is the sample size and k the number of estimated parameters. In the semiparametric case, a corresponding number of degrees of freedom can be approximated. Deviance and (approximate) degrees of freedom of the parametric and the semiparametric model can be used to construct a likelihood ratio test to compare both models (see Buja, Hastie and Tibshirani, 1989; Müller, 1997). The obtained significance levels from these tests are denoted by α. Finally, we listed the pseudo R^2 values, an analog to the linear regression coefficient of determination.

It is obvious to see that models containing variable X5 in the nonparametric part considerably decrease the deviance and increase the coefficient of determination R^2. Accordingly, the significance level for the test of parametric versus nonparametric modeling decreases. In particular, it is below 5% for the both models including X5 alone and including X4, X5 jointly in a nonparametric way.

5.6 Misclassification and Performance Curves

The different fits can be compared by looking at misclassification rates. For the validation, the provided data comprise a subsample (data set C) which was not included in the estimation. We use this validation sample

		Nonparametric in					
	Logit	X2	X3	X4	X5	X4,X5	X2,X4,X5
Deviance	2399.26	2393.16	2395.06	2391.17	2386.97	2381.49	2381.96
df	6118.00	6113.79	6113.45	6113.42	6113.36	6108.56	6107.17
α	–	0.212	0.459	0.130	**0.024**	**0.046**	0.094
pseudo R^2	14.68%	14.89%	14.82%	14.96%	15.11%	15.31%	15.29%

Table 5.4: Statistical characteristics in parametric and semiparametric logit fits. Estimation data set B. Bold values are significant at 5%.

to evaluate all estimators.

The misclassification rates can be pictured by performance curves (Lorenz curves). The performance curve is defined by plotting the probability of observations classified as "good"

$$P(S < s)$$

versus the conditional relative frequency of observations classified as "good" conditioned on "bad"

$$P(S < s|Y = 1).$$

Here, S denotes the score which equals in the parametric logit model

$$S = \sum_{j=2}^{24} \beta_j^T X_j + \beta_0$$

and in the semiparametric logit model

$$S = m_5(X_5) + \sum_{j=2,j\neq 5}^{24} \beta_j X_j$$

when fitting X5 nonparametrically, for instance.

The probability value $P(S < s|Y = 1)$ is a measure for misclassification and thus to be minimized. Hence, one performance curve is to be preferred to another, when it is more downwards shaped.

In practice, the probability $P(S < s)$ is replaced by the relative frequency of classifications $Y = 0$ ("good") given a threshold s. The analog is done for $P(S < s|Y = 1)$. We have computed performance curves for both the estimation data set B and the validation data set C.

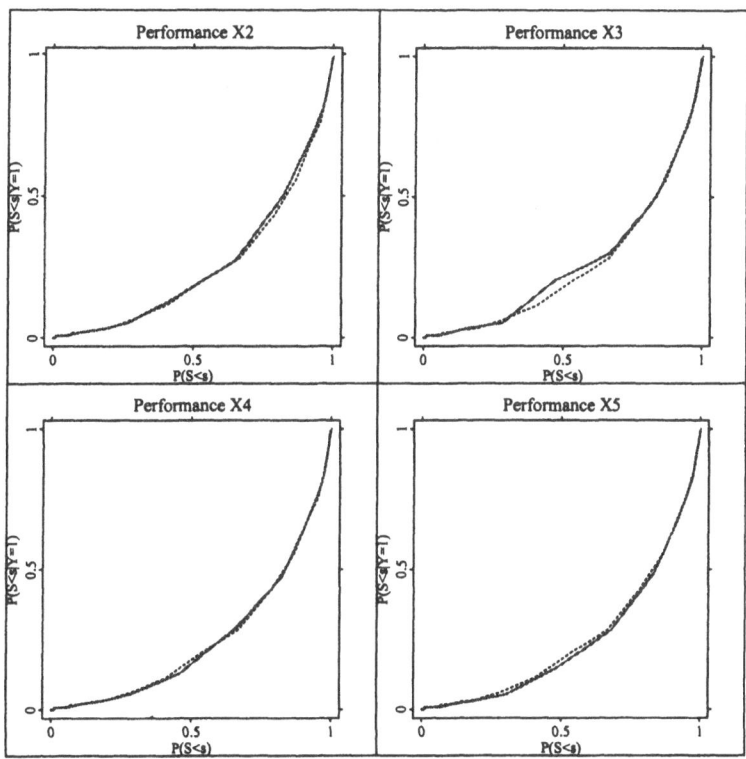

Figure 5.7: Performance curves, parametric logit (black dashed) and semiparametric logit models (thick grey), with variables X2 to X5 (separately) included nonparametrically. Validation data set C.

Figure 5.7 compares the performance of the parametric logit fit and the semiparametric logit fit obtained by separately including X2 to X5 nonparametrically. Indeed, the semiparametric model for the influence of X5 improves the performance with respect to the parametric model. The semiparametric models for the influence of X2 to X4 do not improve the performance with respect to the parametric model, though.

Figure 5.8 compares the performance of the parametric logit fit and the semiparametric logit fit obtained by jointly including X4, X5 nonparametrically. This performance curve improves versus nonparametrically fitting only X4, but shows less power versus fitting only X5. Hence, the improvement of using both variables jointly may be explained by the

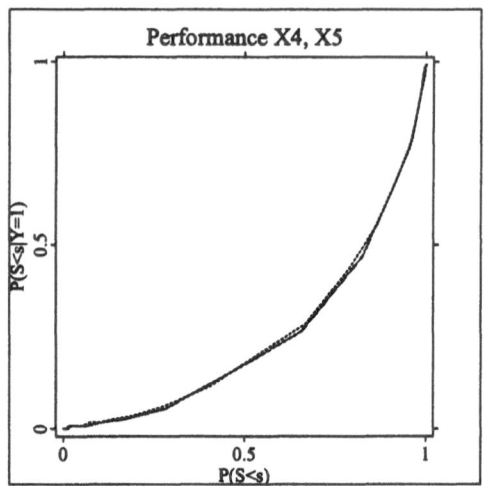

Figure 5.8: Performance curves, parametric logit (black dashed) and semiparametric logit model (thick grey), with variables X4, X5 (jointly) included nonparametrically. Validation data set C.

influence of X5 only.

Bibliography

Arminger, G., Enache, D. and Bonne, T. (1997). Analyzing credit risk data: A comparison of logistic discrimination, classification tree analysis, and feedforward networks, *Computational Statistics, Special Issue: 10 Years AG GLM* **12**: 293–310.

Buja, A., Hastie, T. and Tibshirani, R. (1989). Linear smoothers and additive models (with discussion), *Annals of Statistics* **17**: 453–555.

Gourieroux, C. (1994). Credit scoring. Unpublished script.

Hand, D. J. and Henley, W. E. (1997). Statistical classification methods in consumer credit scoring: a review, *Journal of the Royal Statistical Society, Series A* **160**: 523–541.

Härdle, W., Mammen, E. and Müller, M. (1998). Testing parametric versus semiparametric modelling in generalized linear models, *Journal of the American Statistical Association* **93**: 1461–1474.

Henley, W. E. and Hand, D. J. (1996). A k-nearest-neighbor classifier for assessing consumer credit risk, *Statistician* **45**: 77–95.

McCullagh, P. and Nelder, J. A. (1989). *Generalized Linear Models*, Vol. 37 of *Monographs on Statistics and Applied Probability*, 2 edn, Chapman and Hall, London.

Müller, M. (1997). Computer–assisted generalized partial linear models, *Proceedings of Interface '97*, Vol. 29/1, Houston, TX, May 14–17, 1997, pp. 221–230.

Müller, M., Rönz, B. and Härdle, W. (1997). Computerassisted semiparametric generalized linear models, *Computational Statistics, Special Issue: 10 Years AG GLM* **12**: 153–172.

Severini, T. A. and Staniswalis, J. G. (1994). Quasi-likelihood estimation in semiparametric models, *Journal of the American Statistical Association* **89**: 501–511.

6 On the (Ir)Relevancy of Value-at-Risk Regulation

Phornchanok J. Cumperayot, Jon Danielsson,
Bjorn N. Jorgensen and Caspar G. de Vries

6.1 Introduction

The measurement and practical implementation of the Value–at–Risk (VaR) criterion is an active and exciting area of research, with numerous recent contributions. This research has almost exclusively been concerned with the accuracy of various estimation techniques and risk measures. Compared to the statistical approach, the financial economic analysis of VaR has been relatively neglected. Guthoff, Pfingsten and Wolf (1996) consider the ranking of projects and traditional performance criteria, while Kupiec and O'Brien (1997) and Steinherr (1998) discuss incentive compatible regulation schemes. The wider issue of the benefits for society of VaR based risk management and supervision has hardly been addressed, see however Danielsson, Jorgensen and de Vries (1999b) and Danielsson, Jorgensen and de Vries (1999a). They consider the implications of externally imposed VaR constraints, the public relevance of the VaR based management and regulation schemes, and incentives for quality improvement. This paper summarizes the public policy aspects of this broader line of reasoning.

It is, first of all, important to understand the relation of the VaR measure to other risk measures. The VaR measure is reasonable if interpreted according to its stated intensions, i.e. when it is evaluated truly in the tail area. As demonstrated by Jorion (1998) account of the Long Term Capital Management crisis, VaR risk management is about the tail events. Evaluating VaR deep in the tail area yields good information about infrequent but extreme events about which one should not stay ignorant. Note the emphasis on extreme tail events is counter to most stated industry practice. In addition, the VaR criterion and related alternatives like expected shortfall, have their limitations due to presence of securities with nonlinear payoffs. VaR, as the sole risk objective, may distort bank actions towards excessive risk taking if it permits mangers to become overly focused on expected returns, as noted by Jorion (1998). Since a single minded focus on VaR and expected returns appears too simple minded, we feel that VaR should be modelled as a side constraint on expected utility maximization when structuring portfolios. This has the added benefit that the public relevance of VaR regulation and supervision is better captured by such an interpretation of risk management. When modelled as a side constraint, the VaR restriction can be viewed as a means to internalize externalities of individual risk taking to the financial system at large. To address this issue Danielsson et al. (1999a) discuss how the VaR side constraint affects the equilibrium allocation in a complete markets setting.

The complete markets setting necessarily results in VaR management by financial intermediaries being of limited relevance. We therefore investigate the signals of the VaR measure in incomplete markets. If security markets are incomplete, VaR requirements may correct for market failures. The VaR measure in an incomplete markets setting is unfortunately a double edged sword. Due to the second best nature, the imposition of a crude and sub–optimal VaR constraint may perversely increase the risk in the financial system. From the related literature on solvency regulation we know that crude regulation can enhance risk taking rather than reducing it, see Kim and Santomero (1988) and Rochet (1992). Danielsson et al. (1999b) discuss two cases of incomplete markets and VaR constraints, and demonstrate the possibility of adverse outcomes due to moral hazard and differences in attitudes towards risk taking in combination with limited knowledge on part of the regulators. Taking the limited information of supervisory agencies as given, the potential negative effects of crude VaR based regulation should ideally be minor in comparison to an easily identifiable overriding and glaring market failure to warrant the centralized regulation. We are, however, not aware of such an important market failure and suspect other motives play a role as well.

Using a public choice framework, we suggest that the drive for VaR regulation derives from the regulatory capture by the financial industry to safeguard its monopoly power and the preference of regulators for silent action instead of overt actions like bail outs. We also discuss decentralized alternatives to system wide regulation that may be better at coping with the market imperfections, and are such that these provide positive incentives for quality improvements in the risk management of market participants.

This paper is organized as follows. In section 2 we identify sufficient conditions under which minimizing VaR provides the same unambiguous signal as expected utility maximization and as Expected Shortfall, an alternative risk measure. Further, we discuss the implementation of expected utility maximization subject to attaining a VaR criterion. Section 3 analyzes the effects of introducing regulation based on VaR constraints in an incomplete markets setting. VaR regulation can have an effect even if the confidence level is not binding in the absence of VaR regulation. In section 4, we take a public choice perspective on the drive for VaR regulation. Section 5 summarizes the paper.

6.2 VaR and other Risk Measures

The VaR criterion has three distinct users: internal risk management, external supervisory agencies, and financial statement users (VaR as a performance measure). Although financial institutions are the most common users of VaR, a few non–financial corporations have started to use VaR and provide voluntary disclosures. We investigate conditions under which a VaR objective rank orders projects analogous to other risk criteria. The VaR measure computes the loss quantile q, such that the probability of a loss on the portfolio return R equal to or larger than q is below a prespecified low confidence level, δ:

$$P\{R \leq q\} \leq \delta. \qquad (6.1)$$

If the confidence level, δ, is chosen sufficiently low, the VaR measure explicitly focuses risk managers and regulators attention to infrequent but potentially catastrophic extreme losses. An advantage of the VaR measure is that it may be computed without full knowledge of the distribution of returns R. That is, semi–parametric or fully non–parametric simulation methods may suffice.

In this section, we focus on the relation between the VaR concept and other risk measures in complete markets. We first discuss the established relation between VaR and utility theory. Danielsson et al. (1999a) consider risky projects that can be ranked by Second Order Stochastic Dominance (SOSD). They demonstrate that at sufficiently low quantiles, minimization of VaR provides the same ranking of projects as other risk measures, e.g. the First Lower Partial Moment, the Sharpe Ratio, and Expected Shortfall if the distributions are heavy tailed. Here we provide an alternative argument.

We start by ranking projects by expected utility. Under some conditions on the stochastic nature of projects, projects can be ranked regardless the specific form of the utility functions. To this end, we now introduce the SOSD concept as is standard in financial economics, see Ingersoll

(1987), p. 123, or Huang and Litzenberger (1988).

Definition 6.2.1 *Consider two portfolios i and j whose random returns R_i and R_j are described by cumulative distribution functions F_i and F_j respectively. Portfolio i dominates j in the sense of Second Order Stochastic Dominance (SOSD) if and only if*

$$\int_{-\infty}^{t} F_j(x)dx \geqslant \int_{-\infty}^{t} F_i(x)dx \ \forall t \tag{6.2}$$

or, equivalently,

$$E\left[U(R_i)\right] \geq E\left[U(R_j)\right], \forall U \in \mathcal{U}$$

where

$$\mathcal{U} = \{U : \mathbb{R} \to \mathbb{R} | U \in C^2, U'(.) > 0, U''(.) < 0\}.$$

Suppose the relevant risky prospects can be ordered by the SOSD criterion, and assume that R_i strictly SOSD the other prospects. Define the first crossing quantile q_c as the quantile for which $F_i(q_c) = F_j(q_c)$, for $x \leq q_c : F_i(x) \leq F_j(x)$, and for some $\varepsilon > 0$, $x\varepsilon(q_c - \varepsilon, q_c) : F_i(x) < F_j(x)$. By the definition of SOSD, such a crossing quantile q_c exists, if it is unbounded First Order Stochastic Dominance applies as well. From the definition of the VaR measure (6.1) it is clear that for any $\delta \leq F_i(q_c)$ the VaR quantile q_j from $P\{R_j \leq q_j\} \leq \delta$ is such that $q_j \leq q_i$. Hence at or below the probability level $F_i(q_c)$, the VaR loss level for the expected utility higher ranked portfolio is below the VaR of the inferior portfolio. Minimization of VaR leads to the same hedging decision as maximizing arbitrary utility functions. The important proviso is that the confidence level, δ, is chosen sufficiently conservative in the VaR computations, but this comes naturally with the concept of VaR.

6.2.1   VaR and Other Risk Measures

First Lower Partial Moment

There are other measures that explicitly focus on the risk of loss, we consider two closely related alternatives. The First Lower Partial Moment (FLPM) is defined as:

$$\int_{-\infty}^{t} (t - x)f(x)dx. \tag{6.3}$$

The FLPM preserves the SOSD ranking regardless the choice of the threshold t since, if the first moment is bounded,

$$\int_{-\infty}^{t} (t - x)f(x)dx = \int_{-\infty}^{t} F(x)dx.$$

As was pointed out by Guthoff et al. (1996), it immediately follows that the FLPM and VaR measures provide the same ranking given that δ is chosen sufficiently conservative. Furthermore, they note that the VaR is just the inverse of the zero'th lower partial moment.

Expected Shortfall

Closely related to the FLPM is the Expected Shortfall (ES) measure discussed in Artzner, Delbaen, Eber and Heath (1998) and Artzner, Delbaen, Eber and Heath (1999). The ES measure is defined as

$$ES = \int_{-\infty}^{t} x \frac{f(x)}{F(t)} dx. \tag{6.4}$$

If the definition of SOSD applies, we now show that at conservatively low risk levels, the ES and the VaR measures also coincide.

Proposition 6.2.1 *Suppose the first moment is bounded and that distribution functions are continuous and can be rank ordered by SOSD. Below the first crossing quantile q_c, as defined above, $ES_i \geq ES_j$.*

Proof: If the first moment is bounded, from Danielsson et al. (1999a)

$$ES = t - \frac{FLPM(t)}{F(t)}.$$

At $\delta = F_j(t_{\delta,j}) = F_i(t_{\delta,i}) \leq F_i(q_c)$, we can thus rewrite $ES_i \gtreqless ES_j$ as

$$t_{\delta,i} - \frac{1}{\delta} \int_{-\infty}^{t_{\delta,i}} F_i(x)dx \gtreqless t_{\delta,j} - \frac{1}{\delta} \int_{-\infty}^{t_{\delta,j}} F_j(x)dx$$

where $q_c \geq t_{\delta,i} \geq t_{\delta,j}$. Rearrange the terms as follows

$$\delta(t_{\delta,i} - t_{\delta,j}) \gtreqless \int_{-\infty}^{t_{\delta,i}} F_i(x)dx - \int_{-\infty}^{t_{\delta,j}} F_j(x)dx.$$

Consider the RHS with F_j fixed, but F_i variable. Given $t_{\delta,i}$ and $t_{\delta,j}$, find

$$\sup_{F_i} \left(\int_{-\infty}^{t_{\delta,i}} F_i(x)dx - \int_{-\infty}^{t_{\delta,j}} F_j(x)dx \right).$$

Under the Definition 6.2.1 and the definition of q_c, the admissable F_i are

$$F_j(x) \geq F_i(x) \; \forall x \leq q_c, \text{ and } F_i(t_{\delta,i}) = \delta.$$

Define the following $\widetilde{F}_i(x)$

$$\widetilde{F}_i(x) = \left\{ \begin{array}{l} F_j(x) \text{ for } x \in \; (-\infty, t_{\delta,j}] \\ \delta \text{ for } x \in \; [t_{\delta,j}, t_{\delta,i}] \\ F_i(x) \text{ on } [t_{\delta,i}, \infty), \end{array} \right.$$

Note that $\sup F_i(x) = \widetilde{F}_i(x)$, for $x \in (-\infty, t_{\delta,i}]$. From integration

$$\int_{-\infty}^{t_{\delta,i}} \widetilde{F}_i(x) dx = \int_{-\infty}^{t_{\delta,j}} \widetilde{F}_i(x) dx + \int_{t_{\delta,j}}^{t_{\delta,i}} \widetilde{F}_i(x) dx$$

$$= \int_{-\infty}^{t_{\delta,j}} F_j(x) dx + \delta(t_{\delta,i} - t_{\delta,j}).$$

Thus

$$\sup_{F_i} \left(\int_{-\infty}^{t_{\delta,i}} F_i(x) dx - \int_{-\infty}^{t_{\delta,j}} F_j(x) dx \right) = \delta(t_{\delta,i} - t_{\delta,j}).$$

Hence the RHS at most equals the LHS, since for any other admissable F_i the RHS will be smaller.   □

Artzner et al. (1998) and Artzner et al. (1999) have leveled a critique against the VaR concept on the grounds that it fails to be subadditive. Subadditivity of a risk measure might be considered a desirable property for a risk measure, since otherwise different departments within a bank would not be able to cancel offsetting positions. However, this criticism may not be applicable in the area where VaR is relevant since the above theorem demonstrates that sufficiently far in the tail of the distribution (below the quantile q_c) Expected Shortfall and VaR provide the same ranking of projects. Although, the criticism is applicable if current industry practice of choosing confidence levels at 95 to 99 percent, which is insufficiently conservative. The criticism does not apply when very conservative confidence levels are chosen.

Nevertheless downside risk measures have their problems as well. If an organization becomes too focused on meeting a downside risk objective in combination with large bonuses for traders, it may lead to the following. Initially dealers ensure that the downside risk objective is met through buying the appropriate hedges. Subsequently, dealers use the remaining capital to maximize expected returns, say by buying call options which are far out of the money (which supposedly have the highest expected returns). The resulting kinked payoff function imitates a gambling policy such that a high return occurs with small probability, whereas a low return has a very high probability of occurrence. Dert and Oldenkamp (1997) have dubbed this the casino effect. This effect stems from the implicit objective function within the organization that applies loss aversion as the concept of risk, but displays risk neutrality above the loss threshold. A remedy for this casino effect explored in the next subsection is to view the dealers' problem as maximizing expected utility subject to a side constraint that a given VaR level must be maintained.

6.2.2   VaR as a Side Constraint

In this subsection, we model VaR regulation as a side constraint, where the VaR restriction can be viewed as a means to internalize externalities of individual risk taking to the financial system at large. Regulatory bodies often cite the stability of the payment system as the prime motive for requiring commercial banks to satisfy certain VaR criteria. Hence it seems that the public relevance of VaR regulation and supervision would be better captured by this interpretation.

Grossman and Vila (1989) find that optimal portfolio insurance can be implemented by a simple put option strategy, whereby puts with the strike at the desired insurance level are added to the portfolio, and the remaining budget is allocated as in the unconstrained problem solved at the lower wealth level. Similarly, at the very lowest risk level, the optimal VaR–risk management involves buying a single put option with exercise price equal to the VaR level. But for higher δ–levels, Danielsson et al. (1999a) demonstrate that the optimal VaR strategy generally requires a cocktail of options. Consider a complete market of contingent claims in discrete space. States can then be ordered in keeping with marginal utilities. Suppose the VaR constraint is just binding for the m-th ordered state. The optimal policy then requires an marginal increase in the consumption of the m-th state, at the expense of all other states. Thus optimal VaR management typically, lowers the payoffs of the lowest states, in contrast to the portfolio insurance strategy. The VaR strategy can be effected through buying supershares[1] of that state and to allocate the remaining budget to the solution of the VaR unconstrained problem. In a somewhat different setting Ahn, Boudoukh, Richardson and Whitelaw (1999) consider risk management when op-

---

[1]Supershares are a combination of short and long positions in call options with strikes at the neighboring states

tion prices follow from the Black–Scholes formula. In their model, a firm minimizes VaR with a single options contract subject to a cash constraint for hedging. They derive the optimal options contract for VaR manipulation, and show that a change in funds allocated to the hedging activity only affects the quantity purchased of the put with that particular exercise price.

A benefit of considering complete or dynamically complete capital markets is that no–arbitrage arguments specify the price of all possible hedging instruments independent of risk preferences. The flip side is that risk management is relatively unimportant.[2] A firm can affect its VaR by implementing a hedging strategy; but its capital structure decision as well as its hedging decisions do not affect the Value of the Firm (VoF) in utility terms. Since private agents in a complete markets world can undo any financial structuring of the firm, VaR management by financial intermediaries has only limited relevance, and financial regulation appears superfluous.

6.3   Economic Motives for VaR Management

Incomplete markets are interesting because in such a setting VaR requirements might correct for market failures. To explore this issue, we initially present two examples of market failures from Danielsson et al. (1999b). We then proceed to produce an example of how VaR regulation, that is never binding in the absence of regulation, can still have negative impact on the economy.

Relative to complete markets, incomplete market environments are obviously more complex; hedging can no longer be viewed as a separate activity from security pricing since hedging activities influence the pricing of the underlying securities. For the financial industry and supervisors this became reality during the 1987 market crash. In the Example (6.3.1) below we construct a simple case of feedback between secondary markets and primary markets, i.e. where the price of a put option affects the stock price. In this section at several places we consider two agents labelled A and B, who have the following mean–variance utility functions

$$EU_A = M - \alpha V, \tag{6.5}$$
$$EU_B = M - \beta V,$$

where M denotes the mean and V the variance.

---

[2]See, among others, Modigliani and Miller (1958), Stiglitz (1969b), Stiglitz (1969a), Stiglitz (1974), DeMarzo (1988), Grossman and Vila (1989), and Leland (1998)

Example 6.3.1 *For the two agents A and B with utility functions as given in Equation (6.5), let the risk aversion parameter for A be* $\alpha = 1/8$, *and suppose that agent B is risk neutral so that* $\beta = 0$. *Let A own a risky project with payoffs* $(6, 4, 2, 0, -2)$ *depending on the state of nature, and where each state has equal probability* $1/5$. *Since for this project, the mean is* $M = 2$, *and variance* $V = 8$, *it follows that* $EU_A = 1$. *Agent A is considering selling a 50% limited liability stake in the project. The value of this share is equal to writing a call C with strike at 0. The payoff vector for him changes into*

$$(3 + C, 2 + C, 1 + C, 0 + C, -2 + C),$$

with mean $M = 0.8 + C$ *and variance* $V = 74/25$. *He is willing to write the call if he at least makes* $0.8 + C - \frac{1}{8}\frac{74}{25} \geqslant 1$, *i.e. if* $C \geqslant 57/100$. *So the minimum price for the stock is* 0.57. *The risk neutral agent, B, considers this a good deal since she is willing to pay* $(3 + 2 + 1)/5 = 1.2$ *at the most. Now suppose that B offers to sell a put on the issued stock with strike at 3. She requires at the minimum an option premium of:* $(0 + 1 + 2 + 3 + 3)/5 = 1.8$. *Buying the put as well as selling the stock yields A a payoff vector of*

$$(3 + C - P, 2 + C - P + 1, 1 + C - P + 2, 0 + C - P + 3, -2 + C - P + 3).$$

The mean payoff is $M = C - P + 13/5$, *and the variance* $V = 16/25$. *Hence, assuming that* $P = 1.8$, *we find that A is willing to sell the stock if*

$$\frac{13}{5} + C - \frac{9}{5} - \frac{1}{8}\frac{16}{25} = \frac{18}{25} + C \geqslant 1,$$

thus $C \geqslant 0.28$. *The introduction of the put with strike at 3 lowers the minimum acceptable price of the stock from 57 cents to 28 cents. It can also be shown that the introduction of a put with strike at 1 would lower the minimum desired stock price from 57 cents to 42 cents, assuming that the put trades at 40 cents, the minimum premium requested by B.*

The Example shows that the share price is related to the type of the put being sold. Typically, in incomplete markets the pricing of primary and derivative securities occur simultaneously, see Detemple and Selden (1991). As a consequence, hedging and utility maximization are not independent activities. This joint valuation of all securities is driven by the general equilibrium structure of the economy, i.e. individual marginal utilities, the production structure, and the nature of shocks. These structures are likely far from easily analyzed in concert. In view of this example, an important question for future research appears to be the effect of VaR regulation on the valuation of primary securities. Within the limited scope of this paper we will develop two further examples as a proof of the following second best type proposition from Danielsson et al. (1999b).

Proposition 6.3.1 *If markets are incomplete, VaR based regulation may either increase or decrease welfare.*

In incomplete markets the effects of risk regulation follows from the relation between the VaR criterion and the expected utility, EU, via the value–of–the–firm, henceforth VoF. A negative association between the VaR and VoF measures is less interesting since regulators can sit idle as the firm management has an incentive to capture the gains from VaR management. An example of this is the costly state verification setting considered by Froot, Scharfstein and Stein (1993), and the complete markets environment considered in Leland (1998). In the case below, if the association is positive, VaR management has to be imposed, but ipso facto implies Pareto deterioration, see Danielsson et al. (1999b).

Example 6.3.2 *Let the risk aversion parameter of agent A be $\alpha = 4$, and suppose agent B is risk averse with parameter β to be determined below. Agent A owns a risky project with payoff vector $(1/15, -1/15)$ and state probabilities $(1/2, 1/2)$. This project yields A the expected utility*

$$EU_A = 0 - 4/225 = -8/450.$$

Suppose A can buy a put from B with strike at 0 and premium P. This changes the mean return to $M = \frac{1}{2}\frac{1}{15} - P$, while the variance becomes $1/900$. Hence

$$EU_A(P) = 13/450 - P.$$

Thus A is better off as long as $P < 21/450$. Let the benchmark utility for B be $EU_2 = 0$. Clearly, she is willing to sell the put as long as

$$P \geqslant (15 + \beta/2)/450.$$

Suppose she is willing to sell at fair value: $P = (15 + \beta/2)/450$. Three cases emerge depending how risk averse B is:

1. $0 \leq \beta \leq 12$: *A and B are both better off by trading the put. Hence, in this case the VoF and VaR of agent A are negatively correlated and the VaR does not have to imposed.*

2. $12 < \beta \leq 30$: *A does not want to buy the put since this lowers his expected utility, but buying the put would still reduce his VaR at the 50% probability level. In this case the VoF and VaR correlate positively and VaR regulation has to be imposed.*

3. $\beta > 30$: *B is extremely risk averse, so imposing the purchase of a put on A would not only lower his expected utility, but also increase the VaR.*

The table below summarizes the three possible outcomes from the external imposition of VaR regulation from Example 6.3.2. The lesson is that in incomplete markets agents may often have a positive incentive for risk management themselves. Legislature may therefore think twice before imposing risk constraints. Rather, attention should first be given to provide incentives for better risk management activities, rather than constraining the industry.

Level of Risk Aversion	Impact of VaR regulation
low	Not necessary
medium	Decreases both risk and VoF
high	Increases risk and decreases VoF

Differences in risk attitudes may be viewed as a rather strained basis for analyzing the essential features of risk management in financial markets. Asymmetric information among agents and between agents and regulators is more characteristic. In such a setting moral hazard can have important macro effects as we know for example from the S&L crisis in the United States. In such a setting risk regulation may correct for certain externalities, but can also produce adverse outcomes, again due to the second best solution. As is shown below, adverse effects may arise even when the VaR constraint is *never* binding in the unregulated case. The example builds on Baye (1992) analysis of Stackelberg leadership and trade quota.

Example 6.3.3 *Suppose A and B are banks who have to decide over investing abroad. Bank A is the lead bank and B follows. The investment decisions are interdependent. The decision trees in payoff space and utility space are given in Figures 1 and 2. Bank A has to decide between strategy U and D respectively. After observing this action chosen by bank A, bank B decides on its investment strategy through choosing L or S. Nature plays a role in determining the chances and the outcomes in two states of the world, labelled G and B; positive returns occur with probability 0.8, and negative returns have probability of 0.2.*

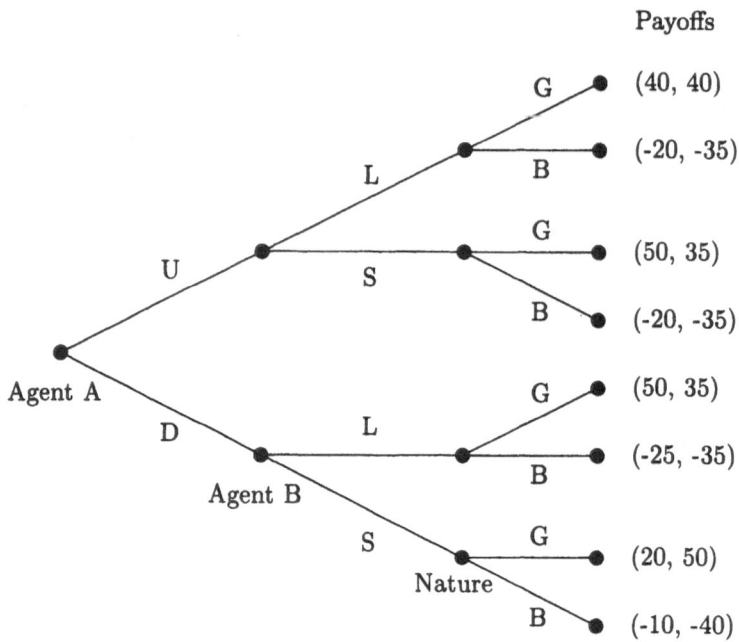

Figure 6.1: Decision tree in payoff space

Assume further that the binomial distributions of returns and the risk aversion parameters are common knowledge among the banks but not to the regulators. In this example we assume that banks have a mean–standard deviation utility function instead of a mean–variance utility function. So replace V in (6.5) by $\sqrt{V}$. The risk aversion parameters are respectively $\alpha = 0.5$ and $\beta = 1.0$. Expected utilities corresponding to each strategy profile are represented in Figure 2.

By backward induction, the strategy combination (U, L) is the only sub-game perfect Nash equilibrium in the unregulated case. Suppose, however, that risk regulation bars banks from losing more than 35. Note such a loss may only occur if the banks would select (D, S), which however is not selected. Nevertheless, such a seemingly innocuous VaR restriction has the effect of changing the equilibrium to (D, L). Since, if bank A selects D, bank B can no longer respond by choosing S. It is then optimal for the lead bank A to switch to D.

Although it is not an intention of the regulators to influence the un-regulated Nash equilibrium, since the restriction is non–binding in that particular equilibrium, it can nevertheless alter the equilibrium. In the

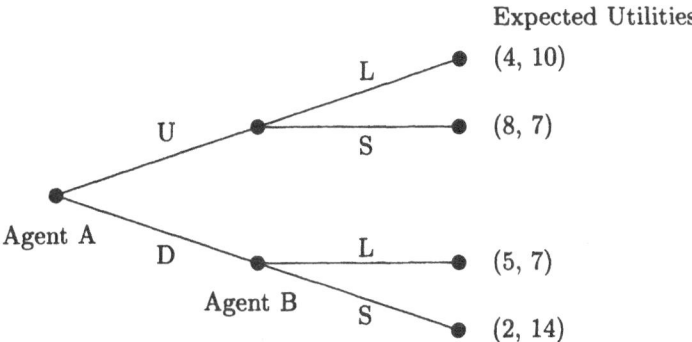

Figure 6.2: Decision tree in utility space

example, regulation obviously induces moral hazard: It provides bank *A* a chance to achieve higher expected utility by bearing the higher risk and meanwhile forcing bank *B* to end up at lower utility. While the VaR of bank *B* is fixed in both equilibria, an increase in the VaR of bank *A* raises social risk and deteriorates social welfare. In fact, the summation of the maximum likely loss of the two banks is minimized in (D, S) and maximized in (D, L). The new equilibrium maximizes systemic risk. Paying too much attention to individual parties' risk, may lead to ignoring the social aspect and consequently systemic risk. Of course we could also have the apparent non–binding VaR regulation produce a reduction in risk in the regulated equilibrium. Suppose that the payoff to bank *A* in the bad state of (D, L) is raised from -25 to -15, then the out off equilibrium VaR constraint barring (D, S) unambiguously reduces the risk in society. This leads to the following result strengthening Proposition 6.3.1.

Proposition 6.3.2 *Even though VaR–based regulation is introduced at a confidence level where it is not binding in the absence of regulation, the introduction of VaR–based regulation may increase overall risk.*

Hence, regulators cannot infer that introducing seemingly non-binding VaR could not be harmful. There are, off course, additional sources of externalities. For example, Danielsson et al. (1999b) also demonstrate that, alternatively, these market failures could be due to a moral hazard with regards to risk management.

6.4   Policy Implications

The previous two sections illustrate that the case for externally imposed risk management rests on strong assumptions about the financial system and that poorly thought out risk control may lead to the perverse outcome of increasing systemic risk. Therefore, any argument in favor of external risk management ought to depend on genuine financial economic arguments. Typically, the case for externally imposed risk management and bank supervision is based on a perceived fear of a systemic failure in the banking industry, in particular the breakdown of the payment clearing system. While it is clear that such failure would be catastrophic, it is less clear that the present regulatory environment is conducive for containment of such a failure. Indeed, as seen above, the current regulatory system may perversely promote systemic risk.

There are numerous episodes of financial crises with sizable failures in the banking sector. Nevertheless, modern history contains few episodes where the clearing system broke down, the classic definition of systemic failure, due to imprudent banking activities. As in many of the bank panics during the era of free banking in the USA, fiscal problems at the state level were the root cause. Free banks were required to hold state bonds on the asset side of their balance sheets, but states often defaulted. The fact that we do not have a large record of systemic failures may also be due to preventive regulatory action and other policy measures, like large scale credit provision by monetary authorities in times of crisis. But since the notion of market risk regulation is a very recent phenomena, it is difficult to imagine how VaR regulation may reduce the chance of an event that has hardly ever been realized. In addition financial history contains many episodes where privately led consortia provided the necessary liquidity to troubled institutions. Thus, even without considering the impact of VaR regulations, it is clear that there are private alternatives to traditional systemic risk containment regulations.

The fundamental protection of the clearing system available to most monetary authorities in a world with fiat currencies is the provision of central bank credit in times of crisis. For obvious reasons of moral hazard, monetary authorities only reluctantly assume the Lender–of–Last–Resort (LOLR) role, but we note that some monetary authorities have an explicit policy of providing liquidity when needed, witness the Federal Reserve System actions after the 1987 market crash.

The alternative to the LOLR role is explicit regulation. Such regulation can take multiple forms. It can be preventive as the tier I and tier II capital adequacy requirements and VaR limits. Second it can consist of ex post intervention, such as the government take-over of banks in Scandinavia and the Korean government intervention forcing the take-overs of five illiquid banks by other banks in 1998. In either case, moral hazard plays an obvious role. First, regulation based on selected variables creates a demand for financial instruments that circumvent or relax these restrictions. Second, if the government policy of bailing out financially distressed financial institutions is correctly anticipated, there is an incentive for excessive risk taking.

There are two avenues open to the authorities, LOLR and preventive regulations, and the trade–offs between them have yet to be investigated. Without knowing the balance between these alternative instruments, we suspect the authorities display a preference for regulation over the LOLR instrument. The use of the latter instrument always implies failure on part of the public authorities and this has real political costs to the authorities. Moreover, the LOLR makes public the costs of the bail out. In contrast regulation, even though costly, goes largely undetected by the public. A desire to keep the cost of systemic failure prevention hidden from the public also tilts political preference in favor of regulation. The preference for hidden actions shows up in may other aspects of public life, e.g. in the prevalence of voluntary export restraints over tariffs in the area of foreign trade, in spite of the costs to society. In addition, institutional economics points out that self preservation and accumulation of power may be the primary objective of bureaucracy.[3] Since regulatory prevention depends on the existence of a sizable administration, while the LOLR role can be executed by a small unit within the monetary authority, the authorities may have a preference for the regulatory approach. Finally, it is possible that financial institutions have willingly embraced regulation, if regulation affects industry structure by deterring entry and expediting exit, or if it facilitates collusion.

There are other more market oriented alternatives to VaR based system wide regulation: The precommitment approach advocated by Kupiec and O'Brien (1997) and the use of rating agencies. In the market oriented proposals more attention is given to incentives for information provision, without the necessity for the regulator to have detailed knowledge of the internal risk management systems. Indeed, the market based regulation adopted by New Zealand in 1986 is combined with more extensive *public* disclosure, see Moran (1997) and Parker (1997). One critique of the current set of Basel rules is that these provide adverse incentives to improve internal risk measurement systems. The market oriented approach would provide the positive incentives. On the other

---

[3]See, among many others, Machiavelli (1977), Hall (1986), and Carpenter (1996) and Carpenter (1998) .

hand such systems leave more room for undetected excessive leverage building. Steinherr (1998) proposes to raise capital requirements for OTC positions, so as to stimulate the migration to organized exchanges, where the mark–to–market principle reduces the risk of systemic melt-down, and individual market participants have a strong incentive for proper risk management. Inevitably, VaR based risk management is likely to stay with us, hopefully it will improve due to a regulatory en-vironment that places more emphasis on individual responsibility.

6.5   Conclusion

When investors are risk averse and their investment opportunities can be ranked through the second order stochastic dominance (SOSD) criterion, Danielsson et al. (1999a) show that if the probability of extreme loss is chosen sufficiently low, the VaR concept attains the same ranking as standard risk measures and other loss criteria. This paper extends this result by showing that without the SOSD ranking, the rankings of investment opportunities based on VaR and Expected Shortfall coincide sufficiently far out in the tail. In our view, however, the VaR criterion is best understood as a side constraint to the expected utility maximization problem of a risk averse investor or portfolio manager. This also has the advantage that we can view the VaR constraint as internalizing the externalities from individual risk taking to the financial system at large. At the margin the VaR side constraint affects the equilibrium allocation within the complete markets setting by increasing the payoff in the state where the constraint is just binding, and by lowering the payoffs to all other states. Within complete markets, however, VaR management by financial intermediaries has only limited relevance, since private agents can undo any financial structuring of the firm. Therefore we shifted attention to incomplete market settings.

When security markets are incomplete, VaR requirements may correct for market failures. Due to the second best nature, however, the reg-ulated outcome may be Pareto inferior to the unregulated case. We present a case where VaR regulation while being non–binding at the prevailing optimal portfolios, can still have real effects since it alters the range for strategic manoeuvering; the off–equilibrium nodes are af-fected by the seemingly innocuous regulation and this in turn changes the playing field. In contrast to the intended effect, the regulation ad-versely affects societal risk taking.[4] Of course full knowledge of the

---

[4]This increase in risk is not necessarily a bad thing, if society at large receives non–marketable gains from the high risk, and supposedly innovative, projects. But such a stimulus is an unforeseen outcome of the VaR regulation.

specific situation at hand on part of the regulator would generate a different policy prescription. But the difficulty with incomplete markets is that it places an enormous informational burden on the supervisory agencies to be able to implement the correct policies. The usefulness of regulation is conditional on the information available to the regulation designers. Absent a detailed pool of information to the supervisors, it seems that regulatory intervention can be warranted only if we can identify an overriding unidirectional negative externality which, when corrected, swamps the costs of regulation.

As a result, one can not present an unambiguous case for regulation over lending of last resort as a mechanism for containing systemic failure. By using public choice arguments to analyze the preference for regulation, we surmise that this preference has more to do with political optimization and preservation of market power than systemic risk minimization.

Bibliography

Ahn, D., Boudoukh, J., Richardson, M. and Whitelaw, R. (1999). Optimal risk management using options, *Journal of Finance* 1(54): 359–375.

Artzner, P., Delbaen, F., Eber, J.-M. and Heath, D. (1998). Thinking coherently, *Hedging With Trees: Advances in Pricing and Risk Managing Derivatives*, Vol. 33, Risk Books, pp. 229–232.

Artzner, P., Delbaen, F., Eber, J.-M. and Heath, D. (1999). Coherent measure of risk, *Mathematical Finance* 3(9): 203–228.

Baye, M. (1992). Quotas as commitment in stackelberg trade equilibrium, *Jahrbücher für Nationalökonomie und Statistik* (209): 22–30.

Carpenter, D. (1996). Adaptive signal processing, hierarchy, and budgetary control in federal regulation, *American Political Science Review* (90): 283–302.

Carpenter, D. (1998). Centralization and the corporate metaphor in executive departments, 1880-1928, *Studies in American Political Development* (12): 106–147.

Danielsson, J., Jorgensen, B. and de Vries, C. (1999a). Complete markets and optimal risk management. Working paper.

Danielsson, J., Jorgensen, B. and de Vries, C. (1999b). Risk management and firm value. Working paper.

DeMarzo, P. (1988). An extension of the modigliani-miller theorem to stochastic economies with incomplete markets and interdependent securities, *Journal of Economic Theory* **2**(45): 353–369.

Dert, C. and Oldenkamp, B. (1997). Optimal guaranteed return portfolios and the casino effect, *Technical Report 9704*, Erasmus Center for Financial Research.

Detemple, J. and Selden, L. (1991). A general equilibrium analysis of option and stock market interaction, *International Economic Review* (2): 279–303.

Froot, K., Scharfstein, D. and Stein, J. (1993). Risk management: Coordinating corporate investment and financing policies, *Journal of Finance* **5**(48): 1629–1658.

Grossman, S. and Vila, J.-L. (1989). Portfolio insurance in complete markets: A note, *Journal of Business* **4**(62): 473–476.

Guthoff, A., Pfingsten, A. and Wolf, J. (1996). On the compatibility of value at risk, other risk concepts, and expected utility maximization, *in* C. Hipp (ed.), *Geld, Finanzwirtschaft, Banken und Versicherungen*, University of Karlsruhe, pp. 591–614.

Hall, P. (1986). *Governing the economy: The Politics of State Intervention in Britain and France*, Oxford University Press, New York.

Huang, C. and Litzenberger, R. H. (1988). Foundations for financial economics. New York, NY, North-Holland.

Ingersoll, J. (1987). *Theory of Financial Decision Making*, Rowman & Littlefield, Savage, MD.

Jorion, P. (1999). Risk management lessons from long-term capital management, *Technical report*, University of California at Irvine.

Kim, D. and Santomero, A. (1988). Risk in banking and capital regulation, *Journal of Finance* (62): 1219–1233.

Kupiec, P. and O'Brien, J. (1997). The pre-commitment approach: Using incentives to set market risk capital requirements, *Technical Report 97-14*, Board of Governors of the Federal Reserve System.

Leland, H. (1998). Agency costs, risk management, and capital structure, *Journal of Finance* (53): 1213–1244.

Machiavelli, N. (1977). *The Prince*, Norton. .

Modigliani, F. and Miller, M. (1958). The cost of capital, corporation finance, and the theory of investment: Reply, *American Economic Review* (49): 655–669.

Moran, A. (1997). On the right reform track, *Australian Financial Review* 7(15).

Parker, S. (1997). Macro focus the brash solution to banking ills, *Australian Financial Review* .

Rochet, J. (1992). Capital requirements and the behavior of commercial banks, *European Economic Review* (36): 1137–1178.

Steinherr, A. (1998). *Derivatives, The Wild Beast of Finance*, Wiley.

Stiglitz, J. (1969a). A re-examination of the modigliani-miller theorem, *American Economic Review* 5(59).

Stiglitz, J. (1969b). Theory of innovation: Discussion, *American Economic Review* 2(59): 46–49.

Stiglitz, J. (1974). On the irrelevance of corporate financial policy, *American Economic Review* 6(64): 851–866.

7 Backtesting beyond VaR

Wolfgang Härdle and Gerhard Stahl

7.1 Forecast tasks and VaR Models

With the implementation of Value-at-Risk (VaR) models a new chapter of risk management was opened. Their ultimate goal is to quantify the uncertainty about the amount that may be lost or gained on a portfolio over a given period of time. Most generally, the uncertainty is expressed by a forecast distribution P_{t+1} for period $t+1$ associated with the random variable L_{t+1}, denoting the portfolio's profits and losses (P&L).

In practice, the prediction P_{t+1} is conditioned on an information set at time t and, typically calculated through a plug-in approach, see Dawid (1984). In this case, P_{t+1} is output of a statistical forecast system, here the VaR model, consisting of a (parametric) family of distributions, denoted by $\mathcal{P} = \{P_\theta \mid \theta \in \Theta\}$ together with a prediction rule. Assumed that P_{t+1} belongs to this parametrized family $\mathcal{P}$ the estimates $\hat{\theta}_t$ are calculated by the prediction rule on the basis of a forward rolling data history $\mathcal{H}_t$ of fixed length n (typically $n = 250$ trading days) for all t, i.e.

$$P_{t+1}(\cdot) = P_{\hat{\theta}_t}(\cdot \mid \mathcal{H}_t).$$

One example for $\mathcal{P}$ also pursued in this paper is the RiskMetrics (1996) delta normal framework, i.e., the portfolios considered are assumed to consist of linear (or linearised) instruments and the common distribution of the underlyings' returns $Y \in IR^d$, i.e., the log price changes $Y_{t+1} = logX_{t+1} - logX_t$, is a (conditional) multinormal distribution, $N_d(0, \Sigma_t)$, where Σ_t (resp. and σ_t^2) denotes a conditional variance, i.e., $\mathcal{H}_t$ measurable function.

Consider for simplicity a position of λ_t shares in a single asset (i.e., d $= 1$) whose market value is x_t. The conditional distribution of L_{t+1} for this position with exposure $w_t = \lambda_t x_t$ is (approximately)

$$
\begin{aligned}
\mathcal{L}(L_{t+1} \mid \mathcal{H}_t) &= \mathcal{L}(\lambda_t(X_{t+1} - x_t) \mid \mathcal{H}_t) = \mathcal{L}\left(w_t \frac{X_{t+1} - x_t}{x_t} \mid \mathcal{H}_t\right) \\
&\approx \mathcal{L}(w_t Y_{t+1} \mid \mathcal{H}_t) = N(0, w_t^2 \, \sigma_t^2),
\end{aligned}
$$

where the approximation refers to

$$lnX_{t+1} - lnx_t = \frac{X_{t+1} - x_t}{x_t} + o(X_{t+1} - x_t).$$

The generalization to a portfolio of (linear) assets is straightforward. Let w_t denote a $d-$dimensional exposure vector, i.e., $w_t = (\lambda_t^1 x_t^1, \cdots, \lambda_t^d x_t^d)$. Hence, the distribution of the random variable $w_t^T Y_{t+1}$ belongs to the family

$$\mathcal{P}_{t+1} = \{N(0, \sigma_t^2) : \sigma_t^2 \in [0, \infty)\}, \qquad (7.1)$$

where $\sigma_t^2 = w_t^T \Sigma_t w_t$.

The aim of the VaR analysis is to estimate $\theta = \sigma_t$ and thereby to establish a prediction rule. For L_{t+1} we adopt therefore the following framework:

$$L_{t+1} \;=\; \sigma_t\, Z_{t+1} \qquad (7.2)$$

$$Z_{t+1} \;\overset{iid}{\sim}\; N(0,1) \qquad (7.3)$$

$$\sigma_t^2 \;=\; w_t^T \Sigma_t w_t. \qquad (7.4)$$

For a given $(n \times d)$ data matrix $\mathcal{X}_t = \{y_i\}_{i=t-n+1,\cdots,t.}$ of realisations of the underlying vector of returns with dimension d, two estimators for Σ_t will be considered. The first is a naive estimator, based on a rectangular moving average (RMA)

$$\hat{\Sigma}_t = \frac{1}{n} \mathcal{X}_t^T \mathcal{X}_t. \qquad (7.5)$$

This definition of $\hat{\Sigma}_t$ makes sense since the expectation of Y_t is assumed zero. The second, also recommended by Taylor (1986) to forecast volatility, is built by an exponential weighting scheme (EMA) applied to the data matrix $\tilde{\mathcal{X}}_t = \{ diag(\lambda^d, \lambda^{d-1}, \cdots, \lambda, 1)^{1/2} y_i\}_{i=t-n+1,\cdots,t}$:

$$\hat{\Sigma}_t = (1 - \lambda) \tilde{\mathcal{X}}_t^T \tilde{\mathcal{X}}_t \qquad (7.6)$$

These estimates are plugged-into (7.4) and (7.2), yielding two prediction rules for

$$P_{t+1} \in \mathcal{P} = \{N(0, \sigma_t^2) \mid \sigma_t^2 \in [0, \infty)\}.$$

By their very nature VaR models contribute to several aspects of risk management. Hence, a series of parameters of interest - all derived from P_{t+1} - arise in natural ways. The particular choice is motivated by specific forecast tasks, e.g., driven by external (e.g., regulatory issues) or internal requirements or needs (e.g., VaR-limits, optimisation issues). A very important part of risk management is the implementation of a systematic process for limiting risk. In the light of that task, it is at hand that forecast intervals defined by the $\widehat{VaR}_t$,

$$\widehat{VaR}_t = F_{t+1}^{-1}(\alpha) := \inf\{x \mid F_{t+1}(x) \geq \alpha\},$$

where F_{t+1} denotes the cdf of P_{t+1}, are substantial.

If the main focus is to evaluate the forecast quality of the prediction rule associated to a VaR model, transformations of F_t should be considered, see Dawid (1984), Sellier-Moiseiwitsch (1993) and Crnkovic and Drachman (1996). For a given sequence of prediction-realisation pairs (P_t, l_t) - where l_t denotes a realisation of L_t - the prediction rules works fine if the sample $u = \{u_t\}_{t=1}^k = \{F_t(l_t)\}_{t=1}^k$ looks like an *iid* random sample from $U[0, 1]$. A satisfactory forecast quality is often interpreted as an adequate VaR model. The focus of this paper is to consider the expected shortfall of L_{t+1}, as the parameter of interest and to derive backtesting methods related to this parameter - this will be done in the next section. The expected shortfall - also called tail VaR - is defined by

$$
\begin{aligned}
E(L_{t+1} \mid L_{t+1} > VaR_t) &= E(L_{t+1} \mid L_{t+1} > z_\alpha \, \sigma_t) && (7.7) \\
&= \sigma_t \, E(L_{t+1}/\sigma_t \mid L_{t+1}/\sigma_t > z_\alpha) && (7.8)
\end{aligned}
$$

where z_α is a α-quantile of a standard normal distribution. The motivation to consider this parameter is threefold. Firstly, McAllister and Mingo (1996) worked out the advantage of (7.7) compared to VaR if these parameters are plugged-into the denominator of a risk performance measures, e.g. a Sharpe-ratio or a RAROC (risk-adjusted return - that constitutes the numerator - on capital) numbers which are used to benchmark divisional performance, see Matten (1996) and CorporateMetrics (1999), - the economic motivation. Secondly, Artzner et al. (1997) and Jaschke and Küchler (1999) pointed out that (7.7) can be used as an approximation for the worst conditional expectation which is a coherent risk measure, a conceptual consideration. Thirdly, Leadbetter, Lindgren and Rootzén (1983) emphasized in the context of environmental regulation the need for incorporating the height of exceedances violating regulatory thresholds and critized those methods solely based on counts, neglecting the heights - statistical arguments. The paper is organised as follows. In the next section we present our approach on backtesting using the expected shortfall risk. In section 3 we apply this methodology to real data and visualise the difference betweeen RMA and EMA based VaRs. Section 4 presents the conclusions of this work.

7.2   Backtesting based on the expected shortfall

As pointed out by Baille and Bollerslev (1992), the accuracy of predictive distributions is critically dependent upon the knowledge of the correct (conditional) distribution of the innovations Z_t in (7.2). For given past returns $\mathcal{H}_t = \{y_t, y_{t-1}, \cdots, y_{t-n}\}$, σ_t in (7.4) can be estimated either by (7.5) or (7.6) and then $\mathcal{L}(L_{t+1} \mid \mathcal{H}_t) = N(0, \hat{\sigma}_t^2)$. This motivates to standardize the observations l_t by the predicted STD, $\hat{\sigma}_t$,

$$\frac{l_{t+1}}{\hat{\sigma}_t}$$

and to interpret these as realizations of (7.2).

$$Z_{t+1} = \frac{L_{t+1}}{\sigma_t} \sim N(0,1) \tag{7.9}$$

For a fixed u we get for Z_{t+1} in (7.2)

$$\vartheta = E(Z_{t+1} \mid Z_{t+1} > u) = \frac{\varphi(u)}{1 - \Phi(u)} \tag{7.10}$$

$$\varsigma^2 = Var(Z_{t+1} \mid Z_{t+1} > u) = 1 + u \cdot \vartheta - \vartheta^2 \tag{7.11}$$

where φ, Φ denotes the density, resp. the cdf of a standard normal distributed random variable.

For a given series of standardized forecast distributions and realizations, $(F_{t+1}(\cdot/\hat{\sigma}_t), l_{t+1}/\hat{\sigma}_t)$ we consider (7.9) as parameter of interest. For fixed u, ϑ is estimated by

$$\hat{\vartheta} = \frac{\sum_{t=0}^{n} z_{t+1} \, I(z_{t+1} > u)}{\sum_{t=0}^{n} I(z_{t+1} > u)} \tag{7.12}$$

where z_{t+1} denotes the realizations of the variable (7.2). Inference about the statistical significance of $\hat{\vartheta} - \vartheta$ will be based on the following asymptotic relationship:

$$\sqrt{N(u)} \left(\frac{\hat{\vartheta} - \vartheta}{\hat{\varsigma}} \right) \xrightarrow{\mathcal{L}} N(0,1) \tag{7.13}$$

where $N(u)$ is the (random) number of exceedances over u and $\hat{\vartheta}$ is plugged-into (7.11) yielding an estimate $\hat{\varsigma}$ for ς. The convergence in (7.13) follows from an appropriate version of the CLT for a random number of summands in conjunction with Slutsky's Lemma, see Leadbetter et al. (1983) for details. Under sufficient conditions and properly specified null hypothesis it is straight forward to prove the complete consistency and an asymptotic α-level for a test based on (7.13), see

Witting and Müller-Funk (1995), pp. 236.

Though these asymptotic results are straight forward they should be applied with care. Firstly, because the truncated variables involved have a shape close to an exponential distribution, hence, $\hat{\vartheta}$ will be also skewed for moderate sample sizes, implying that the convergence in (7.13) will be rather slow. Secondly, in the light of the skewness, outliers might occur. In such a case, they will have a strong impact on an inference based on (7.13) because the means in the nominator and in the denominator as well are not robust. The circumstance that the truncated variables' shape is close to an exponential distribution motivates classical tests for an exponential distribution as an alternative to (7.13).

7.3 Backtesting in Action

The Data The prediction-realisation (P_t, l_t) pairs to be analysed are stemming from a real bond portfolio of a German bank that was held fixed over the two years 94 and 95,i.e., $w_t \equiv w$. For that particular (quasi) linear portfolio the assumptions met by (7.2) - (7.4) are reasonable and common practice in the line of RiskMetrics.

The VaR forecasts are based on a history $\mathcal{H}_t$ of 250 trading days and were calculated by two prediction rules for a 99%-level of significance. The first rule applies a RMA, the second is based on an EMA with decay factor $\lambda = 0.94$ as proposed by RiskMetrics to calculate an estimate of $\hat{\Sigma}_t$ different from (7.5).

Remembering the bond crisis in 1994, it is of particular interest to see how these different forecast rules perform under that kind of stress. Their comparison will also highlight those difficulties to be faced with the expected shortfall if it would be applied e.g. in a RAROC framework.

Exploratory Statistics The following analysis is based on two distinctive features in order to judge the difference of the quality of prediction rules by elementary exploratory means: calibration and resolution, see Murphy and Winkler (1987), Dawid (1984) and Sellier-Moiseiwitsch (1993). The exploratory tools are timeplots of prediction- realization pairs (Fig. 1) and indicator variables (Fig. 4) for the exceedances to analyze the resolution and Q-Q-plots of the variable

$$\frac{L_{t+1}}{VaR_t} = \frac{L_{t+1}}{2.33\sigma_t} \tag{7.14}$$

to analyse the calibration (Fig 2, 3). A further motivation to consider variable (7.14) instead of (7.2) is that their realizations greater than

one are just the exceedances of the VaR forecasts. Of course these
realizations are of particular interest. If the predictions are perfect,
the Q-Q-plot is a straight line and the range of the Y-coordinate of
the observations should fill out the interval $[-1, 1]$. Hence, the Q-Q-
plot for (7.14) visualizes not only the calibration but also the height
of exceedances. A comparison of Figure 2 with Figure 3 shows clearly
that EMA predictions are better calibrated than RMA ones. The second
feature, resolution, refers to the *iid* assumption, see Murphy and Winkler
(1987). Clusters in the timeplots of exceedances, Figure 4,

$$(t, I(l_{t+1} > \widehat{VaR}_t)_{t=1}^{260}$$

indicate a serial correlation of exceedances. Again EMA outperforms
RMA.

From Figure 1, we conclude that in 94 (95) 9 (4) exceedances were

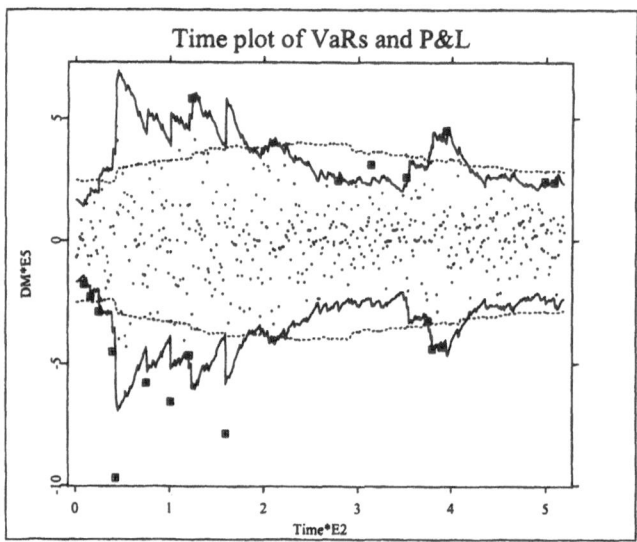

Figure 7.1: The dots show the observed change of the portfolio values,
l_t. The dashed lines show the predicted VaRs based on RMA
(99% and 1%).The solid lines show the same for EMA.

recorded for the EMA and 13 (3) for the RMA. Evidently, the window-
length of 250 days causes an underestimation of risk for RMA if the
market moves from a tranquil regime to a volatile one, and overestimates
vice versa. On the other hand the exponential weighting scheme adapts
changes of that kind much quicker.

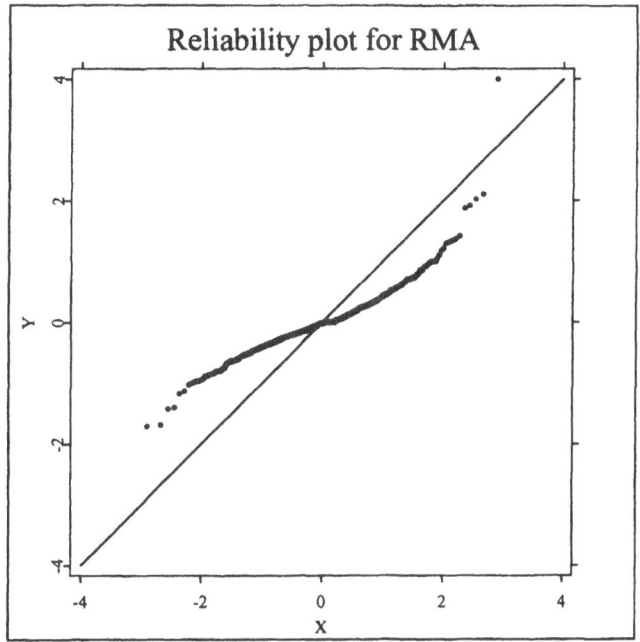

Figure 7.2: Q-Q plot of $l_{t+1}/\widehat{VaR}_t$ for RMA in 94.

The poor forecast performance, especially for the upper tail is evident. The asymmetry and outliers are caused by the market trend. For a particular day the VaR forecast is exceeded by almost 400 %. If the model (7.2) - (7.4) would be correct, the variable (7.14) has a STD of 0.41. The STD calculated from the data is 0.62. Hence, in terms of volatility the RMA underestimates risk on the average of about 50%.

The plot for EMA, Figure 3, shows the same characteristics as those in Figure 2 but the EMA yields a better calibration than RMA. The STD from the data yields 0.5. Hence, an underestimation on the average of 25%. This indicates clearly that EMA gives a better calibration then RMA. Q-Q-plots for 95 are omitted. The two models give similar results, though even in that case the EMA is slightly better.

Inference The exploratory analysis has shown notable differences between the accuracy of RMA and EMA for the year 94. In this paragraph their statistical significance will be investigated. The inference will be

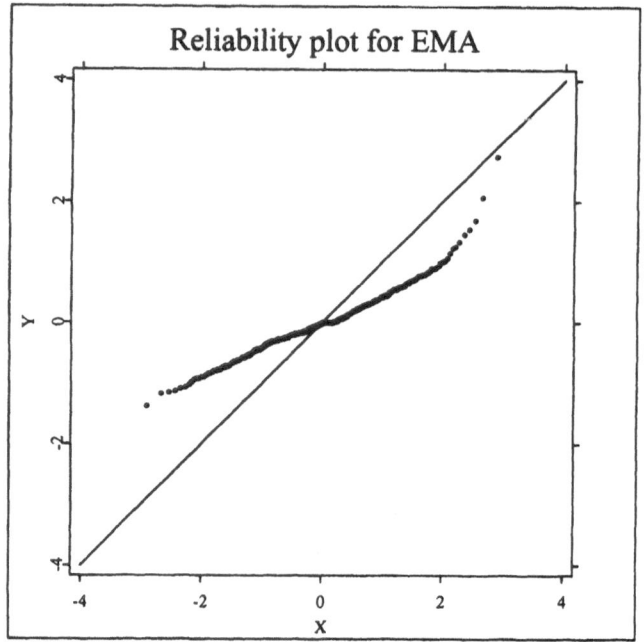

Figure 7.3: Q-Q plot of $l_{t+1}/\widehat{VaR}_t$ for EMA in 94.

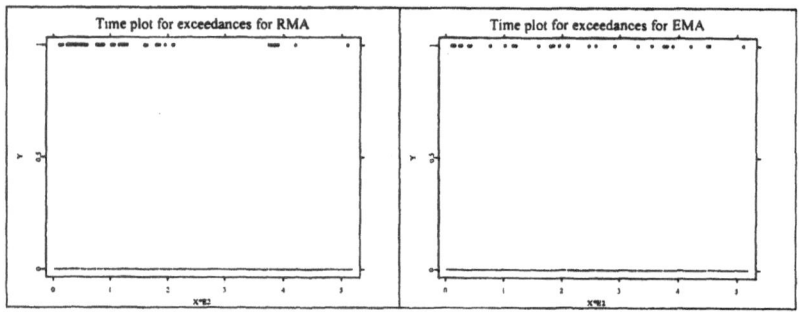

Figure 7.4: Timeplots of the exceedances over VaR of 80% level for RMA (left) and EMA. The better resolution of EMA is evident.

based on the observations

$$\frac{l_{t+1}}{\hat{\sigma}_t}$$

and the underlying model (7.2) - (7.4). The threshold u is set to the 80%-quantile of L_{t+1}/σ_t yielding $\vartheta = 1.4$, by (7.10). Now, based on (7.13) an asymptotic significance test for the hypothesis

$$H_0 : \vartheta \overset{(\leq)}{=} 1.4 \tag{7.15}$$

will be used. This setting - especially (7.2) - seems reasonable for RMA and the given sample of size $n = 250$.

As mentioned by Skouras and Dawid (1996) plug-in forecasting systems have the disadvantage that the uncertainty of the estimator for σ_t is not incorporated in the predictive distribution P_{t+1}. This applies especially to Z_{t+1} if the EMA is used. In that case a $t(n)$-distribution is indicated. A reasonable choice - motivated by generalized degrees of freedom - is

$$Z_{t+1} = \frac{L_{t+1}}{\sigma_t} \sim t(20). \tag{7.16}$$

Though the particular thresholds $u_N = 0.854$ - for the normal distribution - and $u_t = 0.86$ - for the $t(20)$ distribution differ only slightly (0.5 %), the associated means ϑ change about 5 % and the STD ς even about 18%. Parallel to (7.15) the hypothesis

$$H_0 : \vartheta \overset{(\leq)}{=} 1.47 \tag{7.17}$$

will be tested.

Tables 1 to 4 summarize the empirical results.

Method	$\vartheta = 1.4$	$\varsigma = 0.46$	$\frac{\sqrt{N(u)}(\hat{\vartheta}-\vartheta)}{\varsigma}$	significance	nobs
EMA	$\hat{\vartheta} = 1.72$	$\hat{\varsigma} = 1.01$	2.44	0.75%	61
RMA	$\hat{\vartheta} = 1.94$	$\hat{\varsigma} = 1.3$	3.42	0.03%	68

Table 7.1: $H_0 : \vartheta \overset{(\leq)}{=} 1.4$

Method	$\vartheta = 1.47$	$\varsigma = 0.546$	$\frac{\sqrt{N(u)}(\hat{\vartheta}-\vartheta)}{\varsigma}$	significance	nobs
EMA	$\hat{\vartheta} = 1.72$	$\hat{\varsigma} = 1.01$	2.01	2.3%	61
RMA	$\hat{\vartheta} = 1.94$	$\hat{\varsigma} = 1.3$	3.04	0.14%	68

Table 7.2: $H_0 : \vartheta \overset{(\leq)}{=} 1.47$

Firstly from tables 1 and 2, the observed exceedances over threshold u indicate again that the EMA is superior than the RMA. For a sample of

260 prediction-realization pairs 52 exceedances are to be expected (STD 6.45). For the EMA 61 (61 - 52 $\approx$ 1.5 STD) exceedances were observed and 68 (68 - 52 $\approx$ 2.5 STD) for the RMA.

A comparison of table 1 with 2 shows that random errors strongly influence the significance of the test. Recalling the impressive outliers in the Q-Q-plots it is worthwhile to exclude these from the data and re-run the test. The results are given in tables 3 and 4.   Again, a serious

Method	$\vartheta = 1.4$	$\varsigma = 0.46$	$\frac{\sqrt{N(u)}(\hat{\vartheta}-\vartheta)}{\hat{\varsigma}}$	significance	nobs
EMA	$\hat{\vartheta} = 1.645$	$\hat{\varsigma} = 0.82$	2.31	1%	60
RMA	$\hat{\vartheta} = 1.83$	$\hat{\varsigma} = 0.93$	3.78	0.00%	67

Table 7.3: $H_0 : \vartheta \overset{(\leq)}{=} 1.4$ - largest outlier excluded

Method	$\vartheta = 1.47$	$\varsigma = 0.546$	$\frac{\sqrt{N(u)}(\hat{\vartheta}-\vartheta)}{\hat{\varsigma}}$	significance	nobs
EMA	$\hat{\vartheta} = 1.645$	$\hat{\varsigma} = 0.82$	1.65	5%	60
RMA	$\hat{\vartheta} = 1.83$	$\hat{\varsigma} = 0.93$	3.1	0.15%	67

Table 7.4: $H_0 : \vartheta \overset{(\leq)}{=} 1.47$ - largest outlier excluded

change in the level of significance for the RMA is observed indicating the non robustness of the test. These results show furthermore that inference about the tails of a distribution is subtle. In addition the *iid* assumption - cluster of exceedances - might also be violated. One possible source for that is the overlap of the $\mathcal{H}_t$. Hence, the estimates may correlate. Techniques like moving blocks and resampling methods see ? and ? are good remedies.

To overcome the problems related to the slow convergence of (7.13) an exponential distribution may be fitted to the data and then, again a classical test will be applied. The following table reports the significance levels based on a one-sided Kolmogoroff-Smirnov test. Again, the

Method	$\sigma = 0.46$	$\sigma = 0.546$
EMA	0.25%	10% (14%)
RMA	< 0.1%	< 0.1%

Table 7.5: Kolmogoroff-Smirnov Test

results emphasize the impact of random errors. The number in brackets refers to that case, where the largest outlier is deleted.

7.4 Conclusions

VaR models were introduced as specific statistical forecast systems. The backtesting procedure was formulated in terms of measuring forecast quality. The empirical results highlight the better calibration and resolution of VaR forecasts based on (exponentially weights) EMA compared to (uniformly weights) RMA. However, more interesting is the impressive difference in amount (50%). A surprising result is the strong dependence of inferences based on expected shortfall from the underlying distribution. Hence, if expected shortfall will be used in practice in order to calculate performance measures like RAROC the inferences resp. the estimates should be robustified.

Acknowledgements: The authors would like to express their warmest thanks to Zdeněk Hlávka for his continous help by providing the graphics in XploRe. They also wish to thank the Sonderforschungsbereich 373 ("Quantification and Simulation of Economic Processes") for the received financial support. Last but not least the second author disclaims that the views expressed herein should not be construed as being endorsed by the Bundesaufsichtsamt.

Bibliography

Artzner, P., Dealban, F., Eber, F.-J. and Heath, D. (1997) Thinking Coherently, *RISK MAGAZINE*.

Baille, R. T. and T. Bollerslev (1992) Prediction in Dynamic Models with Time-Dependent Conditional Variances. *Econometrica*, **50**: 91–114.

Crnkovic, C. and J. Drachman (1996) A Universal Tool to Discriminate Among Risk Measurement Techniques, *RISK MAGAZINE*.

Dawid, A. P. (1984) The prequential approach. *J. R. Statist. Soc.*, A, **147**: 278–292.

Härdle, W. and Klinke, S. and Müller, M. (1999) XloRe Learning Guide. www.xplore-stat.de, Springer Verlag, Heidelberg.

Jaschke, S. and Küchler, U. (1999) Coherent Risk Measures, Valuation Bounds, and (ν, p)−Portfolio Optimazation. *Discussion Paper*, **No 64**, Sonderforschungsbereich 373 of the Humboldt Universiät zu Berlin

Leadbetter, M. R. (1995) On high level exceedance modeling and tail inference. *Journal of Planning and Inference*, **45**: 247–260.

Matten, C. (1996) *Managing Bank Capital.* John Wiley & Sons: Chicheseter.

McAllister, P. H. and J.J. Mingo (1996) Bank Capital requirements for securitzed loan portfolios. *Journal of Banking and Finance*, **20**: 1381–1405.

Murphy, A. H. and R. L. Winkler (1987) A General Framework for Forecast Verification. *Monthly Weather Review*, **115**: 1330–1338.

RiskMetrics (1996) Technical Dokument, 4th Ed.

CorporateMetrics (1999) Technical Dokument, 1st. Ed.

Sellier-Moiseiwitsch, F. (1993) Sequential Probability Forecasts and the Probability Integral Transform. *Int. Stat. Rev.*, **61**: 395–408.

Skouras, K. and A. P. Dawid (1996) On efficient Probability Forecasting Systems. *Research Report* **No. 159**, Dep. of Statistical Science, University College London.

Taylor, S. J. (1986) *Modelling Financial Time Series.* Wiley, Chichester.

Witting H. and U. Müller-Funk (1995) *Mathematische Statistik II.* Teubner, Stuttgart.

8 Measuring Implied Volatility Surface Risk using Principal Components Analysis

Alpha Sylla and Christophe Villa

8.1 Introduction

The Black-Scholes formula Black and Scholes (1973) (BS hereafter) has remained a valuable tool for practitioners in pricing options as well as a precious benchmark for theoreticians. Indeed, the BS option valuation formula is a one-to-one function of the volatility parameter σ once the underlying stock level S_t , the strike price K and the remaining time to expiration τ are known and fixed. Using the quoted prices of frequently traded option contracts on the same underlier, one can work out the implied volatility $\hat{\sigma}$ by inverting numerically the BS formula. But it is notorious that instead of being constant as assumed by the BS model, implied volatility has a stylized U-shape as it varies across different maturities and strike prices. This pattern called the "smile effect" is the starting point of the implied theories which we concentrate on thereafter.

The basic idea of the implied theories is to price options consistently with the smile drawn from the BS approach. We don't have a view to explaining the smile but to take it as an input for pricing options. In this framework implied volatility can be thought of as the market's assessment of the average volatility over the remaining lifetime of the option contract (see Dupire (1994), Derman and Kani (1994)). Following Dupire (1994), deterministic local volatility functions can be derived from the implied volatility surface thanks to an explicit relation. Having considered only a risk-neutral stochastic process for the underlier, there is a unique diffusion process for the stock that yields such a density conditional on a given point (K, T). After some amount of theory and calculus using arbitrage-free conditions and the Fokker-Planck equation, Dupire (1994) shows that :

$$\tilde{\sigma}^2_{K,T}(t,s) = 2 \left\{ \frac{\frac{\partial C}{\partial T} + \pi K \frac{\partial C}{\partial K}}{K^2 \frac{\partial^2 C}{\partial K^2}} \right\} (t,s)$$

where C is the arbitrage-free price of a European-style call contract with exercise price K and maturity T, π is the instantaneous riskless rate.

But assuming a constant future volatility expectation is also questionable as was shown by Dumas, Fleming and Whaley (1998). In following this empirical fact, Derman and Kani (1998) develop a model which is by many an aspect the Heath, Jarrow and Morton's (HJM) approach to stochastic interest rates. The future evolution of the local volatility surface now depends upon its current state and the movements induced by a certain number of factors. However, the " no free lunch " conditions in the Derman and Kani's framework are considerably involved and cannot be easily implemented. An alternative yet more appealing approach is inspired by the Market Models of the term structure of interest rates addressed by Miltersen, Sandmann and Sondermann (1997) transposed to the current framework by Ledoit and Santa-Clara (1998) and Schönbucher (1999) among others.

This paper has two contributions. The first one is to apply a Principal Components Analysis (PCA) on the implied volatility surface in order to address the issue of determining the number and the nature of the factors that cause the smile to move across time. This is particularly important in smile-consistent derivative pricing. The second point is to use the results in managing the risk of volatility through Value-at-Risk methodology.

8.2   PCA of Implicit Volatility Dynamics

PCA is usually applied to complex systems that depend on a large numbre of factors. The aim is to reduce those factors into a set of composite variables called the Principal Components (PCs) and retain a few of these PCs that explain most of the data variability. PCs are constructed so as not to be correlated over the entire sample and ordered by decreasing " explanatory power ". The purpose of using PCA here is to identify the number and the nature of the shocks that move the Implied Volatility Surface accross time.

For a given maturity bucket, one can run PCA on the daily first differences of implied volatilities corresponding to different moneynesses (Clewlow, Hodges and Skiadopoulos (1998)). Another possibility is to implement PCA on the first differences of implied volatility corresponding to different maturities for a fixed moneyness Avellanda and Zhu (1997).

The availability of at-the-money volatility indexes for most developed stock options markets naturally led us to consider them as inputs in the Market Model of the Term Structure of Implied Volatility insofar as the rest of the implied surface can be valuably approximated thanks to Derman (1999) who pointed out that for strikes not too far from the money :

$$\hat{\sigma}(t, S; T, K) = \hat{\sigma}(t; T) + b(t)(K - S_t),$$

where $\hat{\sigma}(t; T)$ is an at-the-money implied volatility for a maturity T. Therefore, we are going to take at-the-money implied volatility.

8.2.1 Data and Methodology

With respect to CAC 40 index options, we must distinguish between short maturity contracts, PX1 with an American feature and long maturity contracts, PXL designed like an European option. The former has been quoted since November 1988 and the latter since October 1991. For PX1 the longest maturity is six months, whereas that for PXL is two years with fixed expiration in March and September. Based on the following considerations, we use these option prices for our empirical work. First, options written on this index are the one of the most actively traded contracts. Second, data on the daily dividend distributions are available for the index (from the MONEP- SBF CD-ROM). The sample period extends from January 1, 1995 through December 31, 1996. Therefore, the data set includes option prices for 497 trading days. The daily closing quotes for the PXL call and put options are obtained from the MONEP-SBF CD-ROM. The recorded CAC 40 index values are the daily closing index levels.

Following the CBOE methodology we have constructed six CAC 40 volatility indexes corresponding to six maturities of implied volatility : $1, 2, 3, 6, 12$ and 18 months.

8.2.2 The results

The Principal Components Analysis is run on the trading days as individuals and the first differences of log-implied volatility indexes of the maturities considered here as explanatory variables. Unlike Clewlow et al. (1998), we believe that the so-called rules of thumb have proved

Table 8.1: Statistical Summaries of Daily Market Volatility Indexes Level Changes

Statistics	1-month	2-month	3-month	6-month	12-month	18-month
Mean	-0,0124	-0,0134	-0,0126	-0,0118	-0,0108	-0,0099
Std Dev.	1,7698	2,1134	2,3173	0,4980	0,5784	0,6815
Skewness	-0,2961	-0,1798	-0,0341	0,5480	-0,1129	-0,1126
Kurtosis	5,3247	6,7258	5,0319	3,1060	6,2123	6,9574

Table 8.2: PCA results displaying PCs'explanatory power

	Eigenvalues	percentage of data variability explained	cumulative percentage
PC1	2.81	46.85	46.85
PC2	1.89	31.48	78.33
PC3	0.72	12.05	90.38
PC4	0.33	5.45	95.83
PC5	0.23	3.84	99.67
PC6	0.02	0.33	100.00

Table 8.3: PCs' loading on each original variable

	PC1	PC2	PC3
1-month	0.9	-0.21	0.02
2-month	0.97	-0.22	0.01
3-month	0.93	-0.17	0.03
6-month	0.2	0.75	0.55
12-month	0.25	0.86	0.03
18-month	0.29	0.67	-0.64

to be reliable criteria in PCA. Velicer's test is surely another way to cope with the dimensionality reduction problem but it suffers a number of shortcomings which are highlighted by Jolliffe (1989). Furthermore these authors aimed at finding out the risk factors that can be held responsible for moving the local volatility surface (in order to implement Derman and Kani's model) but instead realized their studies on the implied surface, implicitly assuming that both surfaces should necessarily be caused to evolve by the same factors (see discussion thereafter).

Rules of thumb suggest to retain the first three components which explain 90.38% of the total variability contained in the data. The first factor accounts for most part of the data variability and is well correlated with the first three drivers (1, 2, and 3 month volatility indexes).

Their loadings are all above 0.90 (respectively 0.90, 0.97, 0.93) as compared to those of the three remaining volatility indexes (0.20, 0.29, 0.25). This PC is determined mostly by the shorter maturities and reflects a common effect they share. Thus it is remarkable, as " revealed " by previous works, that shorter volatilities are more volatile than longer ones (most of the variance in the sample is due to them because the first PC which represents best their communality has the highest variance).

Moreover a shock on this first factor tends to affect the three shortest volatilities uniformly (as the first PC loadings related to each of them are approximately of the same weight) more than the last three volatilities This PC displays a Z-shape. Conversely a shock on the second factor has little effect on short term volatilities and strongly increases the longer term volatilities. Therefore the second PC also naturally comes out as a Z-shaped factor.

Both factors have symmetric effects which partially compensate so that the resulting effect for a 1% standard deviation shock on each one is an almost less important parallel shift on the volatility surface. Our results are a bit at odds with the well-known direct shift-slope structure highlighted by several studies on the American fixed income security market and related works as well as those of Clewlow et al. (1998) on the volatility skew. It is important to note that the opposite movements of the short term and long term maturity implied volatility along with the first two PCs may explain that an increase in current long term implied volatility (if this also means increase in future short maturity volatility) make the market operators behave in a way which tend to reduce the current short term volatility (by dynamic hedging and feedback effects). This finding illustrates once again the informational role played by derivative markets in providing signals which tend to stabilize or destabilize basis markets. We could also see these movements as the reaction of long term volatilities to short term volatilities. In thinking so we are brought to consider the overreaction hypothesis of long term volatilities to short term volatilities discussed in the literature.

Several authors supported the idea of relatively weak reaction of long term volatilities to shocks affecting short term volatilities in presence of a unit root process in the implied volatilities. On the contrary, some others defended the assumption of overreaction which postulates that long term volatilities react strongly to a shock on short term volatilities. Our remark is that both under-reaction and overreaction may coexist sometimes, or appear only one at a time with roughly the same importance as can be shown by the global PCA (see explanatory power of the PCs). But most often the under-reaction hypothesis is likely to be noticed even in case the two effects would coexist because the under-reaction effect

Table 8.4: PCA results displaying PCs' explanatory power
Short term volatility analysis (1, 2, 3 month terms)

	Eigenvalues	percentage of data variability explained	cumulative percentage
PC1	2.75	91.57	91.57
PC2	0.23	7.74	99.31
PC3	0.02	0.69	100.00

Long term volatility analysis (6, 12, 18 month terms)

	Eigenvalues	percentage of data variability explained	cumulative percentage
PC1	1.95	64.89	64.89
PC2	0.73	24.24	89.13
PC3	0.32	10.87	100.00

linked with the first PC seem to explain more variance than the alternative effect associated with the 2nd PC. The third PC exhibits the fact that a deviation on the third factor induces little effect on the 1, 2, 3 and 12-month volatility indexes. In response to a positive shock, the 6-month volatility increases, the 12-month index remains almost stable whereas the 18-month index undergoes a fall roughly the same amount as the increase in 6-month volatility. PC3 also looks a bit like a Z-shape. It also accounts for 12.05% of the total data variability, a considerable explanatory power as compared with similar studies!. One may rather feel dubious about the shape of this third factor, reason among others which partially justifies the other two PCAs we run afterwards.

Thus we identified three relevant market risk factors which cause the implied volatility surface to move but the " opposite " behaviors of shorter term volatilities on the one hand and longer term volatilities on the other hand invites us to run separate Principal Components Analyses. Our intuition in doing so is that dealing with them separately will bring into light the usual shapes that one naturally expects (shift, slope and curvature). The results are displayed hereafter.

The results fully come up to our expectations. In each analysis the percentage of variance accounted for by the first two PCs rise up to 90% and more (99.31% and 89.13% for short term and long term analysis respectively). Similar conclusions have been reached in other studies in the interest rates and derivative securities literatures. Both analyses come out with an interpretation of the first PC as a shift and the second PC as a slope. The third PC, although a bit marginal (0.69% and 10.87% of data variability explained respectively for short and long term

Table 8.5: PCs' loading on each original variable

	PC1	PC2	PC3
1-month	-0.93	0.37	0.04
2-month	-0.99	-0.06	-0.11
3-month	-0.95	-0.30	0.08
6-month	-0.78	0.56	0.29
12-month	-0.90	0.04	-0.44
18-month	-0.73	-0.64	0.23

analysis), looks like a curvature (notice however the relative importance of the second and third PCs in the long term analysis as compared to other empirical works). No rotation methods are useful here as the PCs fit the data pretty well. The global PCA is rich of teachings but is also a weakness in that it wipes away other important aspects (PCs shapes) that we unveiled by performing separate analyses. Also notice that we recovered three PCs from separate PCA run on three variables, so we are sure that all the variability in the data is contained in those three PCs.

8.3 Smile-consistent pricing models

8.3.1 Local Volatility Models

Local volatility is a close notion to that of implied volatility. We can think of it as a forward volatility meaning that it is the market expectation of future spot volatility at expiry date when market level happens to become equal to exercise price. As time to expiration decreases to zero, an at-the-money [1] option's local volatility tends to instantaneous real volatility In a stochastic local volatility model (see Derman and Kani (1998)), the dynamics of the stock together with its local volatility is described by the following pair of equations :

$$\frac{dS_t}{S_t} = \mu_t dt + \sigma(t) dW_t^0$$

$$d\tilde{\sigma}_{K,T}^2(t, S) = \alpha_{K,T}(t, S) dt + \sum_{i=1}^{n} \theta_{K,T}^i(t, S) dW_t^i$$

---

[1] For sake of brevity, we will use the notations ATM for At-The-Money, OTM for Out-of-The-Money and ITM for In-The-Money all along the paper

The local variance $\tilde{\sigma}^2_{K,T}$ is a risk adjusted expectation of future instantenous variance $\sigma^2(T)$ at time T as

$$\tilde{\sigma}^2_{K,T} = E^{K,T}[\sigma^2(T)]$$

where the expectation $E^{K,T}[...]$ is performed at the present time and market level, and with respect to a new measure which Derman and Kani (1998) called the *K-strike* and *T-maturity forward risk-adjusted measure*.

The spot (instantaneous) volatility at time t, $\sigma(t)$, is the instantaneous local volatility at time t and level S_t, i.e. $\sigma(t) = \tilde{\sigma}_{S_t,t}(t, S_t)$.

All parameters and Brownian motions are one-dimensional and one can usefully report to the authors' paper for accurate discussion over the measurability and integrability conditions of the different parameters, as well as their path and level dependency on past histories of a number of state variables. The second equation specifies a general underlying diffusion process similar to that of Hull & White stochastic volatility models, but instead of assuming an Ito process for the instantaneous volatility parameter, it says that local volatility evolves according to a diffusion process (second equation) in which there are n additional independent sources of risks.

The first component in the second equation is, as in the static case, dependent upon time and stock level whereas the additional Brownian motions account for the stochastic part of the evolution of the volatility surface.

The Market Models of Implied Volatility assume the existence of a sufficient number of traded liquid plain vanilla options. Those ones are used as inputs in addition with the risk-free instantaneous interest rate r and the underlier S under consideration to price exotic options written on that underlier. These prerequisites ensure a complete market since the price of risk is already implicit in the prices of quoted vanilla option contracts. An arbitrage-free condition and a no bubble restriction [2] are derived under which, at expiry date, implied volatility and instantaneous latent volatility are related by a deterministic relation (Schönbucher (1999)).

---

[2] this latter is designed to avoid that implied volatility explode as expiry is approached

8.3.2 Implicit Volatility Models

The prices of the standard liquid option contracts are taken as direct *input* to the model. There is no need to specify the process of the price of risk since it is implicitly provided by the quoted prices. Let us assume that the stock price can be represented this way:

$$dS_t = rS_tdt + \sigma_tS_tdW_t^0$$

where σ_t is a stochastic volatility. The drift r of the process is chosen in such a way so as to ensure that the discounted stock price becomes a martingale under $\mathcal{Q}$.

Each quoted option contract has an implied volatility $\hat{\sigma}_t(K_m, T_m)$. As pointed out by Schönbucher, implied volatility $\hat{\sigma}$ is neither equal to the instantaneous real volatility nor strictly identical to the expected future volatility although the two notions should be close. At expiry date the option pay-off is expressed as:

$$C(S_T, K, T, \hat{\sigma}_T(K, T), r) = \max(S_T - K, 0)$$

and we assume that $\hat{\sigma}_T(K, T) = 0$.

Ledoit and Santa-Clara (1998) provide the proof that an ATM implied volatility converges to the real instantaneous volatility as remaining maturity goes down to zero :

$$\lim_{T \to t} \hat{\sigma}_t(S_t, T) = \sigma_t$$

Observed implied volatilities are of course not constant, instead they evolve stochastically. One can therefore posit a SDE to specify the dynamic process of the implied volatility of an option contract with strike K and maturity T in the following way :

$$d\hat{\sigma}_t(K, T) = u_t(K, T)dt + \rho_t(K, T)dW_t^0 + \sum_{n=1}^{N} v_{n,t}(K, T)dW_t^n$$

The N random sources $(W_1, \cdots, W_N)$ affect the implied volatility dynamics in addition with the leverage term $\rho(K, T)dW_t^0$.

We assume there is a strike-wise continuum of traded options on the same underlier S. As in the HJM multi-factor model of the term structure of interest rates (1992), the model is over-specified insofar as there are more traded assets than the number of random sources. So we will also have to impose restrictions on the dynamics of the drift and volatility parameters to ensure the martingale property of discounted prices.

Schönbucher (1999) relates real instantaneous volatility at $t = T$, and shows by that means the same results as Ledoit and Santa-Clara (1998) for an ATM implied volatility. However Schönbucher's framework brings improvements over the latter by addressing the "volatility bubbles effect" which gives more rigorous arguments to the proof given by Ledoit and Santa-Clara (1998) as to the fact that an ATM implied volatility converges towards to the real instantaneous volatility as $t \to T$. This is so because implied volatility does not explode to infinity once the no bubble restriction holds. It also extends the result to OTM and ITM implied volatilities.

The next section aims at uncovering the random sources that may cause the implied volatility surface to move stochastically through time. The results can be used to gain insight into the determination of the stochastic factors of the local volatility surface, but have then to be considered with great care. We will discuss this issue hereafter.

8.3.3   The volatility models implementation

In light of the empirical results exhibited above in the Principal Components Analyses, we hint a possible enrichment of the two smile-consistent pricing models presented in this paper. But before we use these results, we have to indulge in a brief discussion about the factors of the local volatility surface.

What about the factors of the local volatility surface ?

The implied volatility at time t of the "at-the-money CAC40 option contract expiring at time T" is close to the average of the local volatilities over the remaining lifespan of the contract as assessed by the market at time t (the same relation holds between the yield to maturity and the forward rates). This remark implies that the factors responsible for the local volatility surface at-the-money should be looked for, in the first place, among the factors of the implied surface found out in the Principal Components Analysis. It doesn't mean that those factors have the same interpretation for both surfaces. For instance a slope for the implied surface may stand as a shift factor for the local surface or anything else. All we can say is summed up by a phrase due to Derman and Kani (1998) : " The standard option prices are functionals of the local volatilities at time t and level S just as bond prices are functionals of the forward rates at time t. As a result the dynamical variations of the local volatility surface induce corresponding variations of the standard option prices (i.e the implied volatility surface) ". No mention is made here of a " precise correspondence between the implied and the local factors apart from " one-way induced effects from local to implied ".

The rest of this paper will implicitly assume that the factors of the implied surface are also factors for the local surface because, we know from other studies on the forward rates curve that there are typically two or three factors which account for most of the data variability (remember the fact that in the HJM model the same factors are meant to drive the yield to maturity and the forward rates curves). But we won't venture too far to interpret those factors the same way.

A simple version of the Stochastic Local Volatility model

Derman and Kani propose a " more realistic model " which is specified like this :

$$\frac{dS_t}{S_t} = \mu_t dt + \sigma(t) dW_t^0$$

$$\frac{d\tilde{\sigma}_{K,T}^2(t,S)}{\tilde{\sigma}_{K,T}^2(t,S)} = \alpha_{K,T}(t,S)dt + \theta^1 dW_t^1 + \theta^2 e^{-\lambda(T-t)} dW_t^2 + \theta^3 e^{-\eta(K-S)} dW_t^3$$

Where the spot (instantaneous) volatility at time t, $\sigma(t)$, is the instantaneous local volatility at time t and level S_t, i.e. $\sigma(t) = \tilde{\sigma}_{S_t,t}(t,S_t)$.

The three Brownian motions standing respectively for the parallel shift, the term-structure slope effect, and the skew slope effect of the local volatility surface.

The stochastic differential equation (SDE) becomes in our framework (where we deal with at-the-money volatilities) :

$$\frac{d\tilde{\sigma}_T^2(t)}{\tilde{\sigma}_T^2(t)} = \alpha_T(t)dt + \theta^1 dW_t^1 + \theta^2 e^{-\lambda(T-t)} dW_t^2$$

where the brownian considered here are the shift and slope factors of the implied surface and must not necessarily be interpreted as parallel shift and slope for the local volatility surface. All we suspect is that they are good candidates for implementing the model.

Our previous analyses convinced us that dealing separately with short-term and long-term volatilities is essential in that we can clearly identify the parallel shift-slope structure while exploiting the additional information on the " possible correlation " between short and long term volatilities detected in the global PCA. We propose a more plausible specification which is a system of two simultaneous equations, each for short term and long term volatilities respectively :

$$\frac{d\tilde{\sigma}^2_{T,s}(t)}{\tilde{\sigma}^2_{T,s}(t)} = \alpha_{T,s}(t)dt + \theta^1_s dW^1_{t,s} + \theta^2_s e^{-\lambda(T-t)} dW^2_{t,s}$$

$$\frac{d\tilde{\sigma}^2_{T,l}(t)}{\tilde{\sigma}^2_{T,l}(t)} = \alpha_{T,l}(t)dt + \theta^1_l dW^1_{t,l} + \theta^2_l e^{-\lambda(T-t)} dW^2_{t,l}$$

where the instantaneous relative changes in the short term local volatilities are meant to be driven by the corresponding Brownians. We make further assumptions inspired by the previous Principal Components Analyses:

$$cov(dW^1_{t,s}, dW^1_{t,l}) = \rho^1 dt$$
$$cov(dW^2_{t,s}, dW^2_{t,l}) = \rho^2 dt$$

As one can see, the necessity to distinguish between short term and long term volatilities is essential in the implementation of the model. The idea is not merely theoretical but is also supported by the results brought by Principal Components Analyses and other studies tending to show the reaction of long term volatilities to short term ones (and vice-versa). We presented a way to enrich the " more realistic Derman and Kani's model ".

This work helped to show the "virtues" of PCA in providing an easy, synthetic and visual tool to deal with complex movements on the volatility surface. We are also aware that improvements to this model could be done by introducing the volatility skew in the analysis among others.

The implementation of Schönbucher's Implied Volatility Model is almost straightforward once we know about the number and the nature of the factors involved in the dynamics of the whole implied volatility surface. In this framework as well, a short and long term structure of the implied volatilities is of the same importance as before.

Another important and practical use of the determination of the implied volatility surface risk factors, apart from implementing the stochastic local volatility model, is in managing the risk of volatility. This use is going to be performed through Value-at-Risk (VaR) methodology. This approach to VaR is an alternative to the usual historical VaR and Monte Carlo VaR, each method having its advantages and drawbacks.

8.4 Measuring Implicit Volatility Risk using VaR

In this part we present a risk management tool in the same spirit as that of fixed income securities risk management altogether with the Value-at-Risk methodology. For further details it would be of great interest to report to Golub and Tilman (1997).

8.4.1 VaR : Origins and definition

The matter is : we need to know the minimum loss that can incur for an investor in a portfolio of assets within 5% of the trading days, for a trading day investment horizon without portfolio rebalancing. The notion of Value at Risk is designed to address this day-to-day portfolio manager's concern. Here we deal with a portfolio of derivative assets and we assume market risk factors are the implied volatilities of different maturity buckets (in fact implied volatility indexes).

VaR is defined for a 5% risk level, if we suppose a normal distribution of the changes in the market values in the portfolio, as :

$$Var = 1.65 \times \sigma \left(\Delta \Pi \right) = 1.65 \times \sigma \left(\sum_{i=1}^{n} \Delta \Pi_i \right)$$

$$Var = 1.65 \times \sqrt{ \sum_{i=1}^{n} \sum_{j=1}^{n} Cov \left(\Delta \Pi_i, \Delta \Pi_j \right) }$$

The first transparent and comprehensive VaR methodology was worked out by J.P Morgan to manage portfolio global risk exposure through its RiskMetrics management tool. Since then Value-at-Risk gained popularity among financial institutions and was recommended by the Bank for International Settlements and the Basle committee in 1996. But the practical simplicity of RiskMetrics is also a flaw which gave rise to a great deal of literature about Value-at-Risk. Most studies dealing with portfolios of zero coupon bonds and option contracts have often considered the movements of the yield curve as the main source of market risk but none or only a few coped satisfactorily with the mismeasurements caused by non-linear option contract pay-offs. However, it is well known that option markets are " volatility markets " (i.e. markets where volatility is itself traded like any other asset hence the necessity to know more about the volatility of the volatility)[3]. So one of the most natural way to address the question is to consider the local volatility surface itself (easy to get thanks to available databases). This approach makes sense insofar as near-the-money option contracts are more sensitive to their volatility parameter than the other parameters . Therefore after a short presentation of the concept of VaR, we will go on showing the use of principal components in computing a VaR formula.

8.4.2   VaR and Principal Components Analysis

VaR can be rewritten as :

$$Var = 1.65 \times \sqrt{\sum_{i=1}^{n} \sum_{j=1}^{n} \Pi_i \Pi_j \frac{\sigma\left(\Delta\Pi_i\right)\sigma\left(\Delta\Pi_j\right)}{\Pi_i \Pi_j} \rho\left(\Delta\Pi_i, \Delta\Pi_j\right)}$$

Given the first order approximation $\sigma\left(\Delta\Pi_i\right) \approx \Pi_i \sigma\left(\frac{\Delta\Pi_i}{\Pi_i}\right)$ for any i provided the returns are small enough, we can write :

$$Var = 1.65 \times \sqrt{\sum_{i=1}^{n} \sum_{j=1}^{n} \Pi_i \Pi_j \sigma_i \sigma_j \rho_{i,j}}$$

where Π_i and Π_j are the respective market values of the portfolio and the ith option contract ($i = 1, \cdots, n$). σ_i is the volatility of the relative changes in the ith option market value Π_i due to the movements of the volatility surface. $\rho_{i,j}$ is the correlation between changes in Π_i and Π_j. This approach is not absurd at a first look if we consider that the investment horizon is a day (most traders open up their positions in the first hours of the trading day and close them down before the end of the day or vice versa). Otherwise a portfolio held until the expiry date either ends up in a non limited gain or in a limited loss induced by the premium. A zero mean would then be unreasonable(the distribution of the option payoff at future time T_0 conditional on the information at time T is a truncated lognormal distribution). In order to account for the total effect of the implied volatility surface, we should consider the whole surface,

---

[3]We can easily imagine volatility forward contracts as it is the case for interest rates forward contracts.

i.e all possible maturities available on the market and/or obtained by interpolation from the existent maturities. For practical matters this is of course a waste of time and energy, as Principal Components Analysis provides us with a technique to reduce the set of potentially explanatory variables to a limited subset of these without much loss of information.

Let us now consider a portfolio of option contracts C_i (not necessarily at-the-money so that the risk due to the skew effect should also be addressed), $i = 1, \cdots, n$ with market value Π. We can write for a change in the portfolio value :

$$\Delta\Pi = \sum_{i=1}^{n} \Delta C_i$$

Further we state that the changes in the value of the portfolio are due to risk factors which are reflected by the implied volatility term structure. So one may say that these changes in the portfolio market value are subject to risk through the individual contracts sensitivity to different maturity implied volatility deviations :

$$\Delta\Pi\left(\Delta\hat{\sigma}(T_j), T_j = 1, \cdots, 18\right) = \sum_{i=1}^{n} \Delta C_i\left(\Delta\hat{\sigma}(T_i)\right)$$

We have seen that those first differences of implied volatility are linear combinations of the PCs. So it is natural to state that :

$$\Delta\Pi\left(\Delta PC_k, k = 1, \cdots, 6\right) = \sum_{i=1}^{n} \Delta C_i\left(\Delta PC_k, k = 1, \cdots, 6\right)$$

A Taylor expansion yields for weak deviations :

$$\Delta C_i \approx \frac{\partial C_i}{\partial\hat{\sigma}(T_i)}\Delta\hat{\sigma}(T_i) + \frac{\partial C_i}{\partial r}\Delta r + \frac{\partial C_i}{\partial t}\Delta t + \frac{\partial C_i}{\partial S}\Delta S + \frac{1}{2}\frac{\partial^2 C_i}{\partial S^2}(\Delta S)^2$$

The sensitivity to the risk-free interest rate over a one-day period can be neglected (in periods of high volatilities in interest rates, one can include the implied volatilities extracted from interest rate derivatives in the Principal Components Analysis). The Taylor series boils down to retaining only the sensitivity with respect to the other terms. So we come out with :

$$\Delta C_i \approx \frac{\partial C_i}{\partial \hat{\sigma}(T_i)} \Delta \hat{\sigma}(T_i) + \frac{\partial C_i}{\partial t} \Delta t + \frac{\partial C_i}{\partial S} \Delta S + \frac{1}{2} \frac{\partial^2 C_i}{\partial S^2} (\Delta S)^2$$

For European-style option contracts on an underlyer paying no dividend, sensitivity to the implied volatility or the so-called the "Vega" is computed as :

$$Vega_i = \frac{\partial C_i}{\partial \hat{\sigma}(T_i)} = S\sqrt{T - t}\,\phi\left(\frac{\ln\frac{S_t}{Ke^{-r(T-t)}}}{\hat{\sigma}(T_i)\sqrt{T-t}} + \frac{1}{2}\hat{\sigma}(T_i)\sqrt{T-t}\right)$$

where $K, T - t, S_t, \phi$ respectively stand for the exercise price, the remaining time to maturity, the price of the underlier at time t and the standard normal density function.

Principal Components Analysis allows us to write each difference of implied volatility as a linear combination of the PCs :

$$\Delta \hat{\sigma}(T_i) = \sum_{k=1}^{6} \psi_{ik} PC_k$$

for some coefficients ψ_{ik} (see Table 8.3 above). Finally, we obtain :

$$\Delta \Pi \approx \sum_{i=1}^{n} \left(\sum_{k=1}^{6} Vega_i \psi_{ik} PC_k + Theta_i \Delta t + Delta_i \Delta S + Gamma_i (\Delta S)^2 \right)$$

PCs are independent of one another over the entire sample. Let us also point out that most traders use a delta-gamma neutral hedging, that is the rule. We find, taking these remarks into consideration and taking only the first three PCs into account :

$$Var = 1.65\Pi \sqrt{\sum_{i=1}^{n}\sum_{k=1}^{3}(Vega_i\psi_{ik})^2\lambda_k + \sum_{i=1}^{n}\sum_{j\neq i}\sum_{k=1}^{3}(Vega_i\psi_{ik})(Vega_j\psi_{jk})\lambda_k}$$

Such coefficients as λ and ψ change across time in the same way as $Vega$ is also time and level-dependent. For a computation purpose, one may take an average value for each $\lambda_{k,t}$ got from $T - 2L$ rolling Principal Components Analyses on a constant window width of $2L + 1$ trading days, T being the total number of trading days. However this VaR computation is a " term structure VaR " as it fails to take the skew effect into account. It expresses the risk exposure of the portfolio over a trading day when the manager does not rebalance his portfolio to ensure a continuous vega hedging.

Furthermore PCs are most likely to display ARCH effects so that a constant variance over time may not be a reliable assumption. Finally let us point out that the famous Basle Comitee"$\sqrt{t}$-rule" designed to compute a t-day VaR from a one day VaR derives from the assumption that the value of the aggregate portfolio under consideration follows a geometric brownian motion in a continous time setting. Although it is well-documented that this has little chance to hold true for single basis assets (stocks, interest rates) and exchange rates at least for short periods (see ARCH effects, high correlation pattern between successive price changes or infinite variance over a very short period of time as in Levy processes), one can argue that normality is a reasonable assumption over a sufficiently long period. But as far as a portfolio is concerned, the question is even more intractable. Thus imagine what the issue can look like if the portfolio is highly non linear with respect to the pay-offs of the individual component assets, as it is the case for a gathering of option contracts. All these shortcomings give incentives to implement a VaR prediction formula that is capable of capturing the essential time series features of the portfolio of derivative assets.

Bibliography

Avellanda, M. and Zhu, Y. (1997).   An E-ARCH model for the term structure of implied volatility of FX options, *Applied Mathematical Finance* (1): 81–100.

Black, F. and Scholes, M. (1973).   The pricing of options and corporate liabilities, *Journal of Political Economy* **81**: 637–654.

Clewlow, L., Hodges, S. and Skiadopoulos, G. (1998).   The dynamics of smile, *Technical report*, Warwick Business School.

Derman, E. (1999).   Regimes of volatility, *Risk* pp. 55–59.

Derman, E. and Kani, I. (1994).   Riding on a smile, *Risk* **7**(2): 32–39.

Derman, E. and Kani, I. (1998).   Stochastic implied trees : Arbitrage pricing with stochastic term and strike structure of volatility, *International Journal of Theoritical and Applied Finance* **1**(1): 61–110.

Dumas, B., Fleming, B. and Whaley, W. (1998).   Implied volatility functions, *Journal of Finance* .

Dupire, B. (1994). Pricing with a smile, *Risk* **7**(2): 229–263.

Golub, B. and Tilman, L. (1997). Measuring yield curve risk using principal component analysis, value at risk, and key rate durations, *The Journal of Portfolio Management* .

Jolliffe, I. (1989). *Principal Component Analysis*, Series in Statistics, Springer.

Ledoit, O. and Santa-Clara, P. (1998). Relative option pricing with stochastic volatility, *Technical report*, Working papers, UCLA.

Miltersen, M., Sandmann, K. and Sondermann, D. (1997). Closed form solutions for term strucutre derivatives with log-normal interest rates, *The Journal of Finance* **52**: 409–430.

Schönbucher, P. (1999). A market model for stochastic implied volatility, *Technical report*, Department of Statistics, Bonn University.

9 Detection and estimation of changes in ARCH processes

Piotr Kokoszka and *Remigijus Leipus*

9.1 Introduction

Most statistical procedures in time series analysis rely on the assumption that the observed sample has been transformed in such a way as to form a stationary sequence. It is then often assumed that such a transformed series can be well approximated by a parametric model whose parameters are to be estimated or hypotheses related to them tested. Before carrying out such inferences it is worthwhile to verify that the transformed series is indeed stationary or, if a specific parametric model is postulated, that the parameters remain constant. A classical statistical problem, which is an extension of a two sample problem to dependent data, is to test if the observations before and after a specified moment of time follow the same model. In this paper we are, however, concerned with a change–point problem in which the time of change is unknown. The task is to test if a change has occurred somewhere in the sample and, if so, to estimate the time of its occurrence. The simplest form of departure from stationarity is a change in mean at some (unknown) point in the sample. This problem has received a great deal of attention, see e.g. Csörgő and Horváth (1997). Financial returns series have, however, typically constant zero mean, but exhibit noticeable and complex changes in the spread of observation commonly referred to as clusters of volatility. ARCH models (Engle (1995), Gouriéroux (1997)) have been shown to be well suited to model such financial data. It might appear that detecting a change in variance in an ARCH model will be rather difficult given that they exhibit the aforementioned clusters of volatility. To understand why this need not be the case, recall that if $\{r_k\}$ is a (stationary) ARCH sequence, the variance Er_k^2 is constant; only the conditional variance is a random. It can therefore be hoped that change–point procedures designed for the inference about the mean and applied to the series of squares, $\{r_k^2\}$, whose mean is the variance of the

r_k, will be useful in detecting and estimating a change–point in the parameters of an ARCH sequence which leads to a change in the variance of the r_k.

In this paper we examine CUSUM type tests and estimators, which are simple and easy to implement. We focus on the theoretical aspects of the problem, but before embarking on the detailed technical exposition, we illustrate the basic idea of the CUSUM change–point estimation. The left panel of Figure 9.1 below shows a simulated realization $\{r_k, \ 1 \le k \le 1000\}$ of an ARCH(1) sequence with a change–point whose conditional variance is $a_1 + br_{k-1}^2$ for $k \le 700$ and $a_2 + br_{k-1}^2$ for $k > 700$. We set $a_1 = 1, a_2 = 1.3, b = 0.1$ and used standard normal innovations ε_k. It is impossible to tell by eye if and where a change in variance occurred. The right panel shows the sequence

$$R_k = \frac{k(N-k)}{N^2} \left(\frac{1}{k} \sum_{j=1}^{k} r_j^2 - \frac{1}{N-k} \sum_{j=k+1}^{N} r_j^2 \right)$$

whose elements are weighted differences between sample means of the squares. The weights were chosen in such a way as to minimize bias if a change–point occurs somewhere in the middle of the sample. The change-point estimator, $\hat{k}$, say, is then the value of k which maximizes R_k. In Figure 9.1, the maximum is attained at $\hat{k} = 699$. If several change–points are suspected, the usual procedure is to devide the sample into two parts, before and after an estimated chang e–point, and to test for the presence of a change–point in each of the subsamples. This procedure was applied, for example, by Horváth, Kokoszka and Steinebach (1999).

In the following we consider a general model of the form

$$X_k = (b_0 + \sum_{j=1}^{\infty} b_j X_{k-j})\xi_k, \tag{9.1}$$

where $b_j \ge 0, j = 0, 1, \ldots,$ and $\{\xi_k, k \in \mathbf{Z}\}$ is a sequence of non-negative random variables.

The general framework leading to model (9.1) was introduced by Robinson (1991) in the context of testing for strong serial correlation and has been subsequently developed by Kokoszka and Leipus (2000) and Giraitis, Kokoszka and Leipus (1999a, 1999b). The X_k in (9.1) can be thought of as non–negative powers $|r_k|^\delta$ of the returns r_k. If $\delta = 2$, (9.1) reduces then to the classical ARCH(∞) model

$$r_k^2 = \sigma_k^2 \varepsilon_k^2, \quad \sigma_k^2 = b_0 + \sum_{j=1}^{\infty} b_j r_{k-j}^2, \tag{9.2}$$

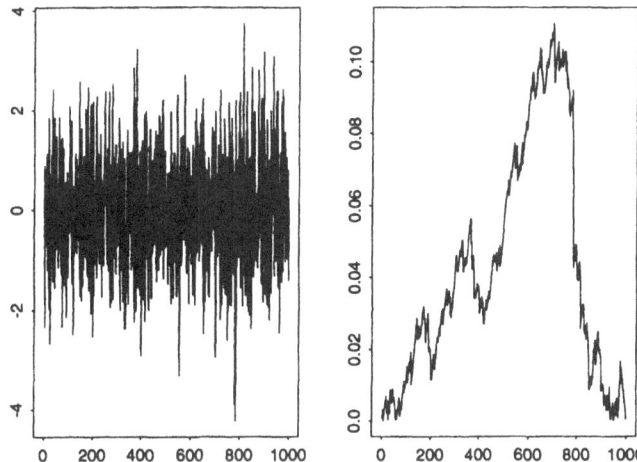

Figure 9.1: A simulated realization of an ARCH(1) sequence with change–point at $k^* = 700$ and the corresponding sequence R_k.

where the ε_k are independent identically distributed random variables with zero mean. The GARCH(p, q) model

$$r_k^2 = \sigma_k^2 \varepsilon_k^2, \quad \sigma_k^2 = \alpha_0 + \sum_{i=1}^{p} \beta_i \sigma_{k-i}^2 + \sum_{j=1}^{q} \alpha_j r_{k-j}^2$$

can be rewritten in form (9.2) under some additional constraints on the coefficients β_i, α_j, see Nelson and Cao (1992).

The interest in studying the powers $|r_k|^\delta$ stems from their complex dependence structure which is often interpreted as long memory, see Baillie, Bollerslev and Mikkelsen (1996), Ding and Granger (1996), Pasquini and Serva (1999), among others. Mikosch and Stărică (1999) seriously challenge such an interpretation and argue that changes in model parameters are a more plausible explanation of the observed characteristics.

In addition to the CUSUM–type procedures described in this paper, (Horváth and Steinebach (2000) view them from a slightly different angle) we are aware of only two other types of change–point procedures for ARCH models. These are Lagrange multiplier tests developed by Chu (1995) and Lundbergh and Teräsvirta (1998) and the procedures of Mikosch and Stărică (1999), which are analogous to the Kolmogorov–

Smirnov and Cramér–von Mises goodness of fit tests in the spectral domain.

We examine tests and estimators based on the process $\{U_N(t), t \in [0,1]\}$, where

$$U_N(t) = \frac{[Nt](N - [Nt])}{N^2} \left(\frac{1}{[Nt]} \sum_{j=1}^{[Nt]} X_j - \frac{1}{N - [Nt]} \sum_{j=[Nt]+1}^{N} X_j \right).$$
(9.3)

The asymptotic theory for the standard statistics like $\sup_{0 \le t \le 1} |U_N(t)|$, $\int_0^1 U_N^2(t) dt$ and their modifications follows automatically from the results presented in Sections 9.2 and 9.3.

The invariance principles for the process $\{U_N(t)\}$ under both the null hypothesis and local alternatives (which converge to the null hypothesis at the rate $1/\sqrt{N}$) are given in Section 9.2. The change-point estimator and its asymptotic properties are discussed in Section 9.3.

9.2   Testing for change-point in ARCH

Denoting by $\mathbf{b} := (b_0, b_1, \dots)$ the parameter sequence in (9.1), we write $\{X_k\} \in \mathcal{R}(\mathbf{b})$ if the X_k satisfy equations (9.1). In applications $\mathbf{b}$ is a function of a finite–dimensional parameter vector.

We consider the null hypothesis

$H_0 : X_1, \dots, X_N$ is a sample from $\{X_k\} \in \mathcal{R}(\mathbf{b})$ for some $\mathbf{b}$

against the change point alternative

$$H_1 : \exists\ \mathbf{b}^{(1)}, \mathbf{b}^{(2)}, \text{ satisfying } \mathbf{b}^{(1)} \neq \mathbf{b}^{(2)},$$

and such that the sample $X_1, \dots, X_N$ has the form

$$X_k = \begin{cases} X_k^{(1)}, & \text{if } 1 \le k \le k^*, \\ X_k^{(2)}, & \text{if } k^* < k \le N, \end{cases}$$

where $\{X_k^{(1)}\} \in \mathcal{R}(\mathbf{b}^{(1)})$, $\{X_k^{(2)}\} \in \mathcal{R}(\mathbf{b}^{(2)})$. As will be seen in the sequel, CUSUM type tests and estimators will work only if the vector $\mathbf{b}$ changes in such a way that the (unconditional) variance of the r_k changes.

The sequences $\{X_k^{(1)}\}$ and $\{X_k^{(2)}\}$ are generated by the same noise sequence $\{\xi_k\}$. Throughout the paper we assume that

$$k^* = [N\tau^*], \quad \text{where } 0 < \tau^* < 1 \text{ is fixed.} \tag{9.4}$$

We write $\overset{D[0,1]}{\longrightarrow}$ to denote the weak convergence in the space $D[0,1]$ endowed with the Skorokhod topology.

9.2.1 Asymptotics under null hypothesis

The results of this subsection follow essentially form a Functional Central Limit Theorem for the process (9.1) obtained by Giraitis, Kokoszka and Leipus (1999) which states that if

$$(E\xi_0^4)^{1/4} \sum_{j=1}^{\infty} b_j < 1 \tag{9.5}$$

then

$$\frac{1}{\sqrt{N}} \sum_{j=1}^{[Nt]} (X_j - EX_j) \overset{D[0,1]}{\longrightarrow} \sigma W(t),$$

where $W(t)$ is a standard Brownian motion and

$$\sigma^2 = \sum_{j=-\infty}^{\infty} \text{Cov}(X_j, X_0). \tag{9.6}$$

Note that if $\xi_k = \varepsilon_k^2$, then condition (9.5) requires, in particular, that $E\varepsilon_k^8 < \infty$. Assumption (9.5) is needed to prove tightness; in order to prove the convergence of finite dimensional distributions it is enough to assume that

$$(E\xi_0^2)^{1/2} \sum_{j=1}^{\infty} b_j < 1. \tag{9.7}$$

Condition (9.7) guarantees that equations (9.1) have a strictly stationary solution such that $EX_k^2 < \infty$, see Giraitis, Kokoszka and Leipus (2000) for the details.

Theorem 9.2.1 *Assume that (9.5) holds. Then under the null hypothesis H_0*

$$U_N(t) \overset{D[0,1]}{\longrightarrow} \sigma W^0(t),$$

where σ^2 is defined by (9.6) and $W^0(t)$ is a Brownian bridge.

For finite order GARCH(p, q) processes, Theorem 9.2.1 can also be deduced from the results of Horváth and Steinebach (2000). This is because such sequences are mixing in a way which implies an appropriate invariance principle in probability.

In order to use Theorem 9.2.1 to construct asymptotic critical regions, we must estimate the parameter σ^2. For this we use the following estimator:

$$\hat{s}_{N,q}^2 = \sum_{|j| \leq q} \omega_j(q)\hat{\gamma}_j, \tag{9.8}$$

where the $\hat{\gamma}_j$ are the sample covariances:

$$\hat{\gamma}_j = \frac{1}{N} \sum_{i=1}^{N-|j|} (X_i - \bar{X})(X_{i+|j|} - \bar{X}), \quad |j| < N, \tag{9.9}$$

$\bar{X}$ is the sample mean $N^{-1} \sum_{j=1}^{N} X_j$ and $\omega_j(q) = 1 - (|j|/q + 1)$, $|j| \leq q$, are the Bartlett weights. Giraitis et al. (1999) show that $\hat{s}_{N,q}$ tends in probability to σ, provided $q \to \infty$ and $q/N \to 0$. Combining this result with Theorem 9.2.1, we obtain the following theorem.

Theorem 9.2.2 *Assume that condition (9.5) is satisfied and $q \to \infty$, $q/N \to 0$. Then under the hypothesis H_0*

$$\frac{U_N(t)}{\hat{s}_{N,q}} \xrightarrow{D[0,1]} W^0(t). \tag{9.10}$$

9.2.2 Asymptotics under local alternatives

In the investigation of statistical properties such as the efficiency of tests, it is important to know the behaviour of the statistics under local alternatives.

Formally, we consider the local alternatives

$$H_1^{(loc)} : \exists\, \mathbf{b}^{(1,N)}, \mathbf{b}^{(2,N)}, \text{ satisfying } \mathbf{b}^{(1,N)} \neq \mathbf{b}^{(2,N)}$$

and

$$X_k^{(N)} = \begin{cases} X_k^{(1,N)} , & \text{if } 1 \leq k \leq k^*, \\ X_k^{(2,N)} , & \text{if } k^* < k \leq N. \end{cases} \tag{9.11}$$

Here $X_k^{(i,N)} \in \mathcal{R}(\mathbf{b}^{(i,N)})$, $i = 1, 2$, with $\mathbf{b}^{(i,N)} \equiv (b_0^{(i,N)}, b_1^{(i,N)}, \dots)$ satisfying the following assumption.

ASSUMPTION A. Assume that for $i = 1, 2$,

$$b_j^{(i,N)} = b_j + \frac{\beta_j^{(i,N)}}{\sqrt{N}}, \quad b_j \geq 0, \ j \geq 0, \tag{9.12}$$

$$\beta_j^{(i,N)} \to \beta_j^{(i)}, \quad \text{as } N \to \infty \tag{9.13}$$

and

$$(E\xi_0^4)^{1/4} \sum_{j=1}^{\infty} (b_j + \beta_j^*) < 1, \tag{9.14}$$

where

$$\beta_j^* := \max_{i=1,2} \sup_{N \geq 1} |\beta_j^{(i,N)}| < \infty. \tag{9.15}$$

Theorems 9.2.3 and (9.2.4) were obtained by Kokoszka and Leipus (1999).

Theorem 9.2.3 *Let* $\lambda := E\xi_0$ *and* $B := \sum_{j=1}^{\infty} b_j$. *Suppose Assumption A holds and set*

$$\sigma^2 = \sum_{k=-\infty}^{\infty} \mathrm{Cov}(Y_k, Y_0), \quad \{Y_j\} \in \mathcal{R}(\mathbf{b}),$$

where $\mathbf{b} = (b_0, b_1, \dots)$, *see (9.12).*

Under the hypothesis $H_1^{(loc)}$

$$U_N(t) \xrightarrow{D[0,1]} \sigma W^0(t) + G(t), \tag{9.16}$$

where

$$G(t) := (t \wedge \tau^* - t\tau^*)\Delta$$

and where

$$\Delta := \frac{\lambda[(\beta_0^{(1)} - \beta_0^{(2)})(1 - \lambda B) + \lambda b_0 \sum_{j=1}^{\infty}(\beta_j^{(1)} - \beta_j^{(2)})]}{(1 - \lambda B)^2}.$$

As pointed out in Subsection 9.2.1, the tests should be based on the rescaled statistic $U_N(\cdot)/\hat{s}_{N,q}$, with $\hat{s}_{N,q}^2$ defined in (9.8). It turns out that under hypothesis $H_1^{(loc)}$, if $q \to \infty$ and $q/\sqrt{N} \to 0$, $\hat{s}_{N,q} \xrightarrow{P} \sigma$, where σ^2 is defined in Theorem 9.2.3. We thus obtain the following theorem.

Theorem 9.2.4 *Under the assumptions of Theorem 9.2.1, if* $q \to \infty$ *and* $q/\sqrt{N} \to 0$,

$$\frac{U_N(t)}{\hat{s}_{N,q}} \xrightarrow{D[0,1]} W^0(t) + \sigma^{-1}G(t),$$

with $\hat{s}_{N,q}^2$ *defined in (9.8).*

Theorem 9.2.4 shows that tests based on the functionals of $U_N(t)/\hat{s}_{N,q}$ have power against local alternatives which converge at the rate $N^{-1/2}$.

9.3   Change-point estimation

Subsection 9.3.1 focuses on ARCH(∞) sequences. Extensions to more general models are discussed in Subsection 9.3.2.

9.3.1   ARCH model

The results presented in this Subsection were obtained in Kokoszka and Leipus (2000).

Assume that the hypothesis H_1 holds, i.e. we observe a sample $X_1, \ldots, X_N$ from the model

$$X_k = \begin{cases} X_k^{(1)}, & \text{if } 1 \leq k \leq k^*, \\ X_k^{(2)}, & \text{if } k^* < k \leq N, \end{cases} \tag{9.17}$$

where

$$X_k^{(i)} = \Big(b_0^{(i)} + \sum_{j=1}^{\infty} b_j^{(i)} X_{k-j}^{(i)}\Big)\xi_k, \quad i = 1, 2. \tag{9.18}$$

Consider a CUSUM type estimator $\hat{k}$ of k^* defined as follows:

$$\hat{k} = \min\{k : |U_k| = \max_{1 \leq j \leq n} |U_j|\}, \tag{9.19}$$

where

$$U_j \equiv U_N(j/N) \text{ with } U_N(t) \text{ given by (9.3)}.$$

It is often convenient to state the results in terms of the estimator of τ^* defined by

$$\hat{\tau} = \hat{k}/N. \tag{9.20}$$

In the following we assume that the $b_j^{(i)}$ decay exponentially fast. More specifically, we suppose that the following assumption holds:

ASSUMPTION B. The coefficients b_j in (9.1) satisfy

$$b_j \leq \beta \alpha^j$$

for some $0 \leq \alpha < 1$ and $\beta \geq 0$ such that

$$E\xi_0^2 \left(\frac{\beta \alpha}{1 - \alpha}\right)^2 < 1.$$

Consistency of the estimator $\hat{\tau}$ follows from the following theorem.

Theorem 9.3.1 *Suppose the sample $X_1, \ldots, X_N$ follows the change–point model defined by (9.17) and (9.18). Consider the change–point estimator $\hat{k}$ given by (9.19). If Assumption B holds and*

$$\Delta := \frac{b_0^{(1)}}{1 - B^{(1)}} - \frac{b_0^{(2)}}{1 - B^{(2)}} \neq 0, \text{ with } B^{(i)} := \sum_{j=1}^{\infty} b_j^{(i)}, i = 1, 2, \quad (9.21)$$

then for any $\epsilon > 0$

$$P\{|\hat{\tau} - \tau^*| > \epsilon\} \leq \frac{C}{\epsilon^2 \Delta^2} N^{-1/2}, \quad (9.22)$$

where C is some positive constant.

The estimator $\hat{\tau}$ remains consistent when the difference $\Delta = \Delta_N$ defined by (9.21) tends to zero, as $N \to \infty$, at a rate slower that $N^{-1/4}$. Formally, suppose that

$$X_k = X_k^{(N)} = \begin{cases} X_k^{(1,N)}, & \text{if } 1 \leq k \leq k^*, \\ X_k^{(2,N)}, & \text{if } k^* < k \leq N, \end{cases} \quad (9.23)$$

where the $X_k^{(i,N)}$, $i = 1, 2$, are given by

$$X_k^{(i,N)} = (b_0^{(i,N)} + \sum_{j=1}^{\infty} b_j^{(i,N)} X_{k-j}^{(i,N)}) \xi_k. \quad (9.24)$$

Denote also $B^{(i,N)} := b_1^{(i,N)} + b_2^{(i,N)} + \ldots$.

Theorem 9.3.2 *Suppose (9.23) and (9.24) hold and Assumption B (with fixed α and β) is satisfied for any $N \geq 1$ and $i = 1, 2$. Assume also that $b_0^{(1,N)}$, $b_0^{(2,N)}$ are bounded sequences and*

$$\Delta_N := b_0^{(1,N)}/(1 - B^{(1,N)}) - b_0^{(2,N)}/(1 - B^{(2,N)})$$

satisfies $\Delta_N \to 0$ and $|\Delta_N| N^{1/4} \to \infty$ as $N \to \infty$. Then $\hat{\tau}$ is a weakly consistent estimator of τ^*.

The difference $\hat{\tau} - \tau^*$ is of the same order as in the case of a least squares estimator studied by Bai (1994) in the case of weakly dependent linear sequences. Specifically, under the assumptions of Theorem 9.2.1

$$|\hat{\tau} - \tau^*| = O_P\left(\frac{1}{N \Delta_N^4}\right). \quad (9.25)$$

9.3.2   Extensions

The approach developed for the estimation of a change-point in ARCH models can be extended to a more general setting. The results presented in this subsection have been obtained in Kokoszka and Leipus (1998). As observed in the introduction, estimation of a change in variance of an ARCH sequence can be reduced to the estimation of a chance in mean of the sequence of squares. In the following we consider a rather general model for a change–point in mean.

Suppose the sample $X_1, \ldots, X_N$ has the form

$$X_k = \begin{cases} X_k^{(1)}, & \text{if } 1 \le k \le k^*, \\ X_k^{(2)}, & \text{if } k^* < k \le N, \end{cases} \tag{9.26}$$

where $X_k^{(1)}$ has mean $\mu^{(1)}$ and $X_k^{(2)}$ has mean $\mu^{(2)}$ with $\Delta := \mu^{(1)} - \mu^{(2)} \ne 0$.

Consider a family of estimators $\hat{k} = \hat{k}(\gamma)$ of k^*, $0 \le \gamma < 1$, defined by

$$\hat{k} = \min\{k : |\tilde{U}_k| = \max_{1 \le j < N} |\tilde{U}_j|\}, \tag{9.27}$$

where

$$\tilde{U}_k = \left(\frac{k(N-k)}{N}\right)^{1-\gamma} \left(\frac{1}{k} \sum_{j=1}^{k} X_j - \frac{1}{N-k} \sum_{j=k+1}^{N} X_j\right).$$

In applications, considering the whole range $0 \le \gamma < 1$ offers the advantage of being able to apply many estimators to a sample under study. The results below are valid under the following assumption.

ASSUMPTION C. The observations $\{X_k, 1 \le k \le N\}$ follow (9.26) and satisfy

$$\sup_{1 \le k \le m \le N} \text{Var} \sum_{j=k}^{m} X_j \le C \, (m - k + 1)^\delta,$$

for some $0 \le \delta < 2$ and $C > 0$.

Theorem 9.3.3 *Suppose Assumption C holds and consider the change-point estimator $\hat{k}$ given by (9.27). Then*

$$P\{|\hat{\tau} - \tau^*| > \epsilon\} \le \frac{C}{\epsilon^2 \Delta^2} \begin{cases} N^{\delta/2-1}, & \text{if } \delta > 4\gamma - 2, \\ N^{\delta/2-1} \log N, & \text{if } \delta = 4\gamma - 2, \\ N^{2\gamma-2}, & \text{if } \delta < 4\gamma - 2. \end{cases}$$

Bai (1994), Antoch, Hušková and Prášková, Z. (1997), Horváth (1997), Horváth and Kokoszka (1997), among others, studied the model (9.26) with $X_k^{(i)} = \mu^{(i)} + Y_k$, $i = 1, 2$, where $\{Y_k\}$ is a zero-mean stationary sequence. The model considered here is more general because the change in mean may be due to some very general change in the structure of a possibly non–linear sequence. For a sample $X_1, \ldots, X_N$ following the ARCH(∞) model (9.26), (9.18) satisfying Assumption B, Assumption C holds with $\delta = 1$. For other examples see Kokoszka and Leipus (1998). Note also that Assumption C is stated in terms of the observable sequence $\{X_k\}$.

Similarly as in Theorem 9.3.2, one can verify that the estimator $\hat{\tau}$ is still consistent if the magnitude of the jump Δ depends on the sample size N and tends to zero, as $N \to \infty$, at a sufficiently slow rate.

Bibliography

Antoch, J., Hušková, M. and Prášková, Z. (1997). Effect of dependence on statistics for determination of change, *Journal of Statistical Planning and Inference* **60**: 291–310.

Bai, J. (1994). Least squares estimation of a shift in linear processes, *Journal of Time Series Analysis* **15**: 453–472.

Baillie, R., Bollerslev, T. and Mikkelsen, H. (1996). Fractionally integrated generalized autoregressive conditional heteroskedasticity, *Journal of Econometrics* **74**: 3–30.

Chu, C.-S. (1995). Detecting parameter shift in GARCH models, *Econometric Reviews* **14**: 241–266.

Csörgő, M. and Horváth, L. (1997). *Limit Theorems in Change-Point Analysis*, Wiley, New York.

Ding, Z. and Granger, C. (1996). Modeling volatility persistence of speculative returns: A new approach, *Journal of Econometrics* **73**: 185–215.

Engle, R. (1995). *ARCH Selected Readings*, Oxford University Press, Oxford.

Giraitis, L., Kokoszka, P. and Leipus, R. (1999). Detection of long memory in ARCH models, *Technical report*, preprint.

Giraitis, L., Kokoszka, P. and Leipus, R. (2000). Stationary ARCH models: dependence structure and Central Limit Theorem, *Econometric Theory* . forthcoming.

Gouriéroux, C. (1997). *ARCH Models and Financial Applications*, Springer.

Horváth, L. (1997). Detection of changes in linear sequences, *Ann. Inst. Statist. Math.* **49**: 271–283.

Horváth, L. and Kokoszka, P. (1997). The effect of long–range dependence on change–point estimators, *Journal of Statistical Planning and Inference* **64**: 57–81.

Horváth, L., Kokoszka, P. and Steinebach, J. (1999). Testing for changes in multivariate dependent observations with applications to temperature changes, *Journal of Multivariate Analysis* **68**: 96–119.

Horváth, L. and Steinebach, J. (2000). Testing for changes in the mean and variance of a stochastic process under weak invariance, *Journal of Statistical Planning and Inference* . forthcoming.

Kokoszka, P. and Leipus, R. (1998). Change–point in the mean of dependent observations, *Statistics and Probability letters* **40**: 385–393.

Kokoszka, P. and Leipus, R. (1999). Testing for parameter changes in ARCH models, *Lithuanian Mathematical Journal* **39**: 231–247.

Kokoszka, P. and Leipus, R. (2000). Change-point estimation in ARCH models, *Bernoulli* **6**: 1–28.

Lundbergh, S. and Teräsvirta, T. (1998). Evaluating GARCH models, *Technical Report No. 292*, Working paper, Stockholm School of Economics.

Mikosch, T. and Stărică, C. (1999). Change of structure in financial time series, long range dependence and the GARCH model, *Technical report*, preprint available at http://www.cs.nl/~eke/iwi/preprints.

Nelson, D. and Cao, C. (1992). Inequality constraints in the univariate GARCH model, *Journal of Business and Economic Statistics* **10**: 229–235.

Pasquini, M. and Serva, M. (1999). Clustering of volatility as a multiscale phenomenon, *Technical report*, preprint.

Robinson, P. (1991). Testing for strong serial correlation and dynamic conditional heteroskedasticity, *Journal of Econometrics* **47**: 67–84.

10 Behaviour of Some Rank Statistics for Detecting Changes

Aleš Slabý

10.1   Introduction

The one who is to explore systematic changes within one's data is put before several questions bound with possible alternatives to the null hypothesis that there is no change of distribution along the data. The questions above all include:

- How many changes should we expect?

- What kinds of changes are likely to appear?

- How to estimate nuisance parameters to be consistent under both the null and the alternative?

- Can we get appropriate critical values for chosen statistics and our data?

These questions are imperious since the choice of alternative is to be tested usually affects the choice of testing statistics in a crucial way. We made an attempt to develop statistics for testing changes in location which are "robust" from this point of view, and above all, tried to provide a plausible way how to obtain reasonable critical values.

The situation of the null hypothesis can be represented by a model where we have independent observations $X_1, \ldots, X_n$, all with the same common continuous distribution function F. As we consider changes in location we make the distribution F depend on some location parameter θ, thus, write $F(x, \theta) = F(x - \theta)$ and consider generally that $X_i \sim F(x - \theta_i)$. Now the null hypothesis can be expressed as

$$X_i = \theta + e_i, \qquad i = 1, \ldots, n, \tag{10.1}$$

where $e_1, \ldots, e_n$ are i.i.d. random variables with the distribution function $F(x, 0)$. The alternative of abrupt change means in these terms that

$$X_i = \theta + \delta I_{\{i>m\}} + e_i, \qquad i = 1, \ldots, n, \tag{10.2}$$

where δ stands for the size of the change appearing at an unknown point $m < n$. Assuming a linear trend, unremitting gradual change can be formulated as

$$X_i = \theta + \delta \frac{i-m}{n} I_{\{i>m\}} + e_i, \qquad i = 1, \ldots, n. \tag{10.3}$$

Here δ represents the speed of the change starting at an unknown point $m < n$. These are only two basic alternatives, later we will consider other, more complicated ones.

The testing statistics in question are based on centered partial sums

$$S_{1k} = \sum_{i=1}^{k} (a(R_i) - \bar{a}_n), \tag{10.4}$$

$$S_{2k} = \sum_{i=k+1}^{n} (a(R_i) - \bar{a}_n) \frac{i-k}{n}, \tag{10.5}$$

where $k = 1, \ldots, n$. $R_1, \ldots, R_n$ are the ranks of observations $X_1, \ldots, X_n$, $a(1), \ldots, a(n)$ are scores and

$$\bar{a}_n = \frac{1}{n} \sum_{i=1}^{n} a(i). \tag{10.6}$$

Also, note that $S_{1n} = 0 = S_{2n}$. The only conditions we impose upon the scores are that there exist finite positive constants D_1, D_2 and η such that

$$\frac{1}{n} \sum_{i=1}^{n} (a(i) - \bar{a}_n)^2 \geq D_1 \tag{10.7}$$

and

$$\frac{1}{n} \sum_{i=1}^{n} |a(i) - \bar{a}_n|^{2+\eta} \leq D_2. \tag{10.8}$$

S_{1k} and S_{2k} are linear rank statistics, for $k = 1, \ldots, n$, it can be easily shown that

$$\text{Var } S_{1k} = \frac{k(n-k)}{n} \sigma_n^2(a), \tag{10.9}$$

$$\text{Var } S_{2k} = \left(\frac{(n-k)(n-k+1)(2(n-k)+1)}{6n^2} - \frac{(n-k)^2(n-k+1)^2}{4n^3} \right) \times$$
$$\times \sigma_n^2(a), \tag{10.10}$$

and

$$\text{Cov } (S_{1k}, S_{2k}) = -\frac{k(n-k)(n-k+1)}{2n^2} \sigma_n^2(a), \tag{10.11}$$

where

$$\sigma_n^2(a) = \frac{1}{n-1} \sum_{i=1}^{n} (a(i) - \bar{a}_n)^2, \tag{10.12}$$

so that there is no need for estimation of variance. This fact allows to avoid obstacles connected with the third question. However, despite S_{1k} and S_{2k} are distribution free under (10.1) it is not true under alternatives and a procedure sensitive to the alternative have to be performed in case of rejection of the null hypothesis. The well known statistic

$$C_1 = \max_{k=1,\dots,n-1} (\text{Var } S_{1k})^{-1/2} |S_{1k}| \tag{10.13}$$

is used for testing abrupt changes (10.2) whereas

$$C_2 = \max_{k=1,\dots,n-1} (\text{Var } S_{2k})^{-1/2} |S_{2k}| \tag{10.14}$$

is proposed to test (10.3). Generally speaking C_1 is more sensitive to abrupt changes, and vice versa, C_2 is more affected by gradual ones. Also, in the next section we realize that the limit behaviour is different. That is why we introduce a testing statistic based on a quadratic form of (S_{1k}, S_{2k}), namely

$$Q = \max_{k=1,\dots,n-2} (S_{1k}, S_{2k})^T [\text{Var } (S_{1k}, S_{2k})]^{-1} (S_{1k}, S_{2k}). \tag{10.15}$$

It is pleasant that the inverse of the variance matrix can be expressed in the following compact explicit form,

$$[\text{Var } (S_{1k}, S_{2k})]^{-1} = \frac{\sigma_n^{-1}(a)}{(n-k-1)} \times \tag{10.16}$$
$$\times \begin{bmatrix} \frac{2n(2(n-k)+1)}{k(n-k)} - \frac{3(n-k+1)}{k} & \frac{6n}{(n-k)} \\ \frac{6n}{(n-k)} & \frac{12n^2}{(n-k)(n-k+1)} \end{bmatrix},$$

which also makes clear why we take maximum over k only up to $n-2$. Note that $S_{1,n-1} = -S_{2,n-1}$.

The statistic (10.15) is sensitive to both alternatives (10.2) and (10.3), and after rejecting the null hypothesis (10.1) by Q one can continue in analysis using the statistics C_1 and C_2. If we omitted the test based on (10.15) and carried out two separate tests by means of (10.13) and (10.14) to ascertain whether abrupt or gradual change is present, we could not control the level of such test since C_1 and C_2 are dependent. From (10.9), (10.10) and (10.11), for instance, we can see that S_{1k} and S_{2k} become highly correlated as k/n increases. The test based on Q treats this problem by mimicking the idea of F-tests. This idea could also be employed to test more general groups of alternatives which include quadratic trend etc.

Another modification can be done by considering differences between moving sums instead of the partial sums (10.4), (10.5) themselves, thus consider

$$S_{1k}^*(G) = S_{1,k+G} - 2S_{1k} + S_{1,k-G} \tag{10.17}$$

and

$$S_{2k}^*(G) = S_{2,k+G} - 2S_{2k} + S_{2,k-G} \tag{10.18}$$

for $k = G + 1, \dots, n - G$. The moving sums sequentially aim at only a part of data (of length $2G$), which allows for testing multiple changes through employing multistage procedures. Such alternatives and proper procedures are discussed in the last section.

We introduce statistics analogous to the above in (10.13) through (10.15) and put

$$M_1(G) = \max_{k=G+1,\dots,n-G} (\operatorname{Var} S_{1k}^*(G))^{-1/2} |S_{1k}^*(G)|, \tag{10.19}$$

$$M_2(G) = \max_{k=G+1,\dots,n-G} (\operatorname{Var} S_{2k}^*(G))^{-1/2} |S_{2k}^*(G)|, \tag{10.20}$$

and

$$Q^*(G) = \max_{k=G+1,\dots,n-G} \left\{ (S_{1k}^*(G), S_{2k}^*(G))^T [\operatorname{Var} (S_{1k}^*(G), S_{2k}^*(G))]^{-1} \times \right.$$

$$\left. \times (S_{1k}^*(G), S_{2k}^*(G)) \right\}. \tag{10.21}$$

Again $S_{1k}^*(G)$, $S_{2k}^*(G)$ are linear rank statistics and straightforward calculations yield

$$\operatorname{Var} S_{1k}^*(G) = 2G \, \sigma_n^2(a), \tag{10.22}$$

$$\operatorname{Var} S_{2k}^*(G) = \left(\frac{G(2G^2 + 1)}{3n^2} - \frac{G^4}{n^3} \right) \sigma_n^2(a), \tag{10.23}$$

and

$$\text{Cov}\,(S_{1k}^*(G), S_{2k}^*(G)) = -\frac{G}{n}\,\sigma_n^2(a). \qquad (10.24)$$

Hence

$$[\text{Var}\,(S_{1k}^*(G), S_{2k}^*(G))]^{-1} = \sigma_n^{-1}(a)\left(\frac{(2G-1)(2G+1)}{3} - \frac{2G^3}{n}\right)^{-1} \times$$

$$\times \begin{bmatrix} \frac{2G^2+1}{3G} - \frac{G^2}{n} & \frac{n}{G} \\ \frac{n}{G} & \frac{2n^2}{G} \end{bmatrix} \qquad (10.25)$$

for arbitrary meaningful G.

10.2   Limit Theorems

Unfortunately, explicit formulae for distributions of the introduced statistics are not known. One way how to get critical values is to employ a limit approximation. The limit behaviour of C_1 and $M_1(G)$ is well known.

Theorem 10.2.1 *Let $X_1, \ldots, X_n$ be i.i.d. random variables with common continuous distribution function F. Let assumptions (10.7) and (10.8) be satisfied.*

1. *As $n \to \infty$*

$$P\left\{\sqrt{2\log\log n}\,C_1 \le y + 2\log\log n + \frac{1}{2}\log\log\log n - \frac{1}{2}\log\pi\right\}$$
$$\longrightarrow \exp(-2e^{-y}) \qquad (10.26)$$

for any real y where C_1 is defined in (10.13).

2. *If moreover*

$$G \longrightarrow \infty \quad and \quad G^{-1}n^{2/(2+\eta)}\log n \longrightarrow \infty, \qquad (10.27)$$

then for any real y, as $n \to \infty$,

$$P\left\{\sqrt{2\log\frac{n}{G}}\,M_1(G) \le y + 2\log\frac{n}{G} + \frac{1}{2}\log\log\frac{n}{G} - \frac{1}{2}\log\frac{4\pi}{9}\right\}$$
$$\longrightarrow \exp(-2e^{-y}), \qquad (10.28)$$

where $M_1(G)$ is defined in (10.19).

See (Hušková, 1997) for the proof. The following theorem demonstrates that the statistics based on (10.5) exhibits different behaviour from many points of view.

Theorem 10.2.2 *Let $X_1, \ldots, X_n$ be i.i.d. random variables with common continuous distribution function F. Let assumptions (10.7) and (10.8) be satisfied.*

1. *As $n \to \infty$*

$$P\left\{\sqrt{2\log\log n}\, C_2 \leq y + 2\log\log n + \log\frac{\sqrt{3}}{4\pi}\right\} \longrightarrow \exp(-2e^{-y})$$

(10.29)

   for arbitrary real y where C_2 is defined in (10.14).

2. *If moreover (10.27) holds then for arbitrary real y, as $n \to \infty$,*

$$P\left\{\sqrt{2\log\frac{n}{G}}\, M_2(G) \leq y + 2\log\frac{n}{G} + \log\frac{\sqrt{3}}{4\pi}\right\} \longrightarrow \exp(-2e^{-y}),$$

(10.30)

   where $M_2(G)$ is defined in (10.20).

The first assertion of Theorem 10.2.2 is a consequence of Theorem 2 in (Jarušková, 1998) and auxiliary Theorem 3 in (Hušková, 1997). Derivation of the second assertion follows similar ideas. However, complete proofs are far beyond the scope of this paper and will be published elsewhere. If we went into the proofs we would realize that the gradual changes are bound with Rayleigh process of order 2 whereas in the abrupt case the order appears to be 1. See also Theorem 12.3.5 in (Leadbetter et al., 1983) or more general Theorem 10.6.1 in (Bergman, 1992). Similar situation arises under alternatives, study (Hušková, 1996; Hušková, 1998) to become more familiar with this topic. It suggests strange behaviour of Q and $Q^*(G)$ and prevents from deriving a limit theorem in a standard way. It is a challenge into the future. We will further discuss it in the next section.

10.3   Simulations

We realized that an elegant, simple and fast enough tool for our purposes can be S-Plus language, and think that mentioning several tricks may appear worth for some readers. For instance, generating a random permutation of order n takes only a line

```
a <- order(runif(n))
```

In our computations we used Wilcoxon scores and in such a case the command above generates just the scores. Otherwise one have to apply appropriate function. After generating scores and subtracting proper mean (10.6) one can immediately obtain a vector $(S_{1,1}, \ldots, S_{1,n-1})$ via

```
S1 <- cumsum(a[1:(n-1)])
```

Computation of $(S_{2,1}, \ldots, S_{2,n-1})$ is more delicate. First refine (10.5) as

$$S_{2k} = \frac{1}{n} \sum_{i=k+1}^{n} \sum_{j=i}^{n} (a(R_i) - \bar{a}_n), \tag{10.31}$$

then according to (10.31)

```
S2 <- rev(cumsum(cumsum(rev(a[2:n])))))/n
```

gives $(S_{2,1}, \ldots, S_{2,n-1})$. This method avoids using ineffective for-cycle and takes approximately only $2n$ arithmetic operations. Having calculated S1 and S2 we are ready to compute vectors $(S_{1,G+1}^*(G), \ldots, S_{1,n-G}^*(G))$ and $(S_{2,G+1}^*(G), \ldots, S_{2,n-G}^*(G))$ as the second order difference of lag G:

```
S1G <- diff(c(S1, 0), lag=G, differences=2)
S2G <- diff(c(S2, 0), lag=G, differences=2)
```

We omit the rest since it is a simple application of (10.9) through (10.16), and (10.19) through (10.25) for computed S1, S2, and S1G, S2G.

The simulated distributions of (10.13), (10.14), (10.19), (10.20), (10.15), and (10.21) are plotted in Figures 10.1 through 10.6 and were obtained from 10000 realizations of random permutations. Respective limit distributions standardized according to (10.26), (10.29), (10.28), (10.30) are drawn as dashed lines in Figures 10.1 through 10.4. In all cases simulated quantiles at levels 90%, 95%, and 99% are outlined.

Figures 10.1 and 10.2 confirm the results of the limit theorems. The convergence is very slow but apparent. The two figures also illustrate the different types of limit behaviour. One can see that approximation by limit quantiles is inadequate for C_1 and respective tests are very conservative. Although the approximation is far better for C_2 it is not necessarily good enough, e.g. taking $n = 100$ the limit 95%-quantile is close to the simulated 99%-quantile. See also Table 10.2.

An interesting comparison to Table 1 in (Jarušková, 1998) can be done. The limit approximation is worse in case of ranks for smaller sizes but the discrepancy appears to decrease whereas in case of i.i.d. normal variables it stagnates. It can be well explained. In fact ranks are not independent but the dependency becomes weak as n increases. Moreover, the limit theorems use approximation by certain Gaussian stationary processes. Also, note that convergence of maxima of normal variables is extremely slow, see (Embrechts, Klüppelberg and Mikosch, 1997) for more information. The described phenomenon can be expected in case of the other introduced statistics too, and the reasoning is the same.

The situation in case of moving sums is complicated by entrance of the bandwidth parameter G. The choice of bandwidth crucially influences convergence properties. The choice $G_1 = 1.5\sqrt{n}$ for $M_1(G_1)$ and $G_2 = \sqrt{n}/1.5$ for $M_2(G_2)$ expresses our experience. Detailed analysis of possible choices would be a study by itself and therefore we omit detailed discussion because of lack of space. In the next section we give some recommendations which involve in consideration other aspects too. Figures 10.3 and 10.4 imply similar conclusions as Figures 10.1 and 10.2. The limit approximation of quantiles is better for $M_2(G_2)$ than for $M_1(G_1)$ again. However, here the convergence of the whole distribution seems to be better for $M_1(G_1)$. Note that relationship between behaviour of partial sums and their second order differences is distinct in Theorems 10.2.1 and 10.2.2, see also first order differences in Theorem 2 of (Hušková, 1997).

Figure 10.5 shows shapes more similar to some Fréchet distribution, which does not have finite moments from certain order. The figures discussed above exhibit better accordance with Gumbel distribution shape. However, if we consider $\sqrt{Q}$ it turns out to be the other way around. Table 10.1 contains Hill estimates for the shape parameter ξ of generalized extreme value distribution, see (Embrechts et al., 1997). The table shows that the estimates for $\sqrt{Q}$ are even closer to 0 than the estimates for both C_1 and C_2. It incites an idea that the convergence of Q could be better. As for $\sqrt{Q^*(G)}$ versus $M_1(G_1)$ and $M_2(G_2)$ it is not true. Note that here we take $G = \sqrt{n}$ as a compromise between G_1 and G_2.

n	C_1	C_2	$M_1(G_1)$	$M_2(G_2)$	$\sqrt{Q}$	$\sqrt{Q^*(G)}$
100	0.066	0.090	0.063	0.049	0.060	0.057
250	0.067	0.084	0.062	0.047	0.058	0.049
1000	0.059	0.083	0.048	0.044	0.045	0.052

Table 10.1: Hill estimates for the shape parameter

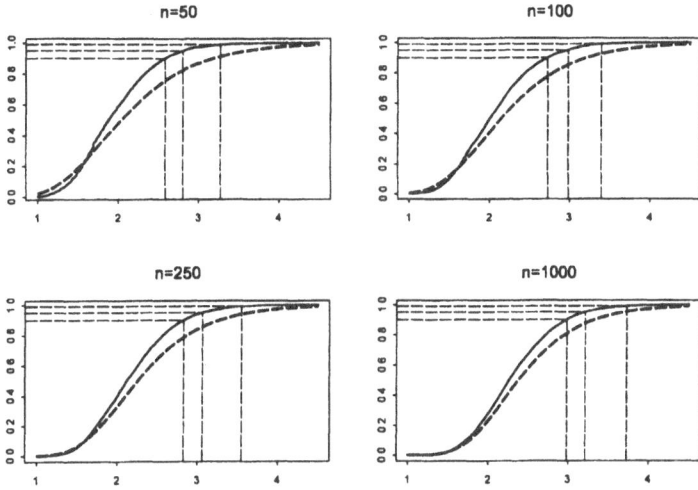

Figure 10.1: Simulated convergence of C_1 (Wilcoxon scores)

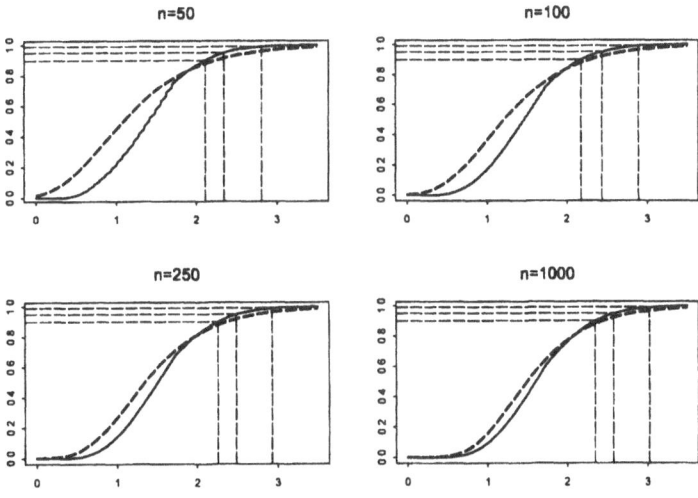

Figure 10.2: Simulated convergence of C_2 (Wilcoxon scores)

10.4 Comments

The statistics (10.19), (10.20), (10.21) are appropriate for testing combined alternatives. The multistage testing procedures consist of repeated

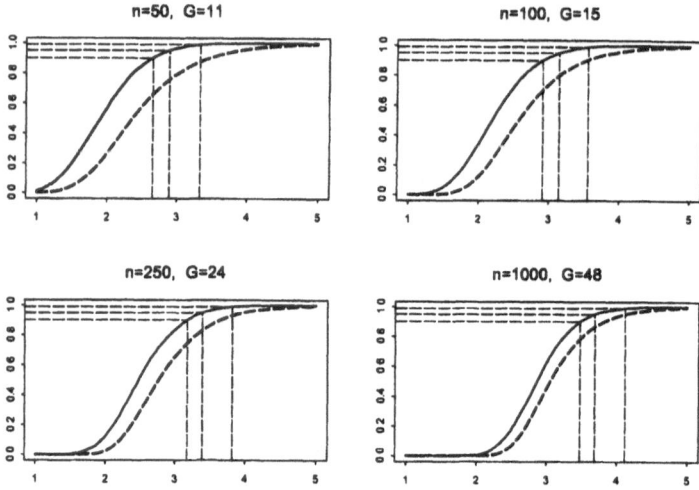

Figure 10.3: Simulated convergence of $M_1(G)$ (Wilcoxon scores)

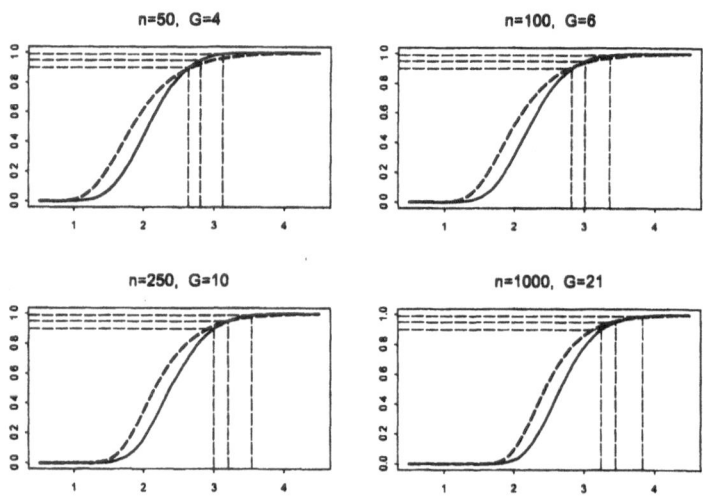

Figure 10.4: Simulated convergence of $M_2(G)$ (Wilcoxon scores)

testing applied to disjunct sections of data. The procedures can be designed recursively as follows. Find the maximum over the whole data and decide whether it is significant. If yes exclude $2G$ observations around the point where the maximum is attained and apply the same

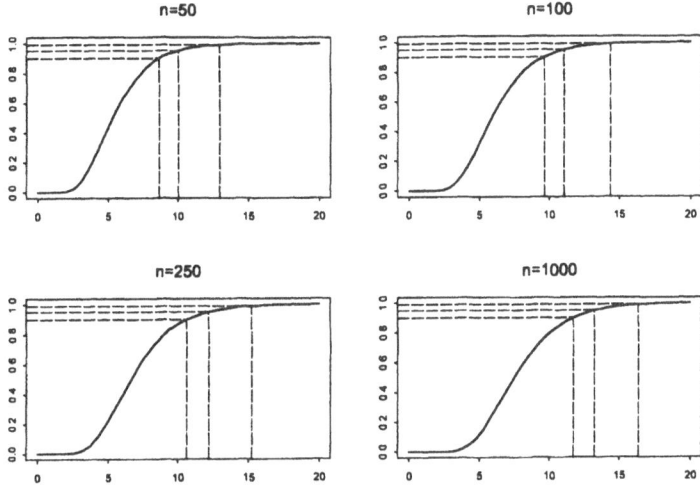

Figure 10.5: Simulated distributions of Q (Wilcoxon scores)

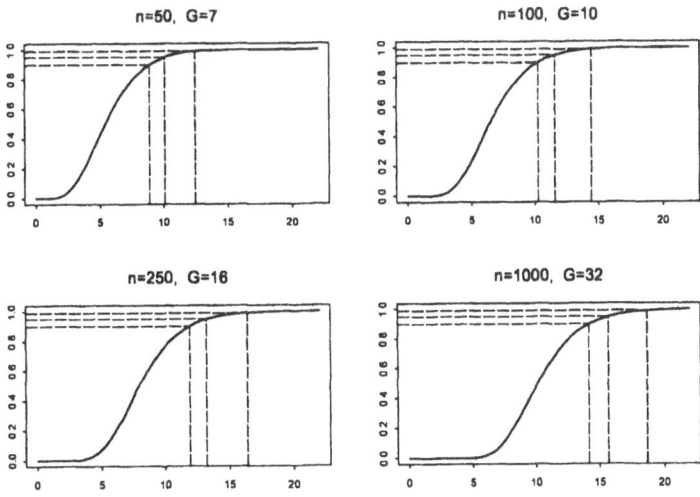

Figure 10.6: Simulated distributions of $Q^*(G)$ (Wilcoxon scores)

step again for the two data sections which arise this way. The procedure ends when tests in all sections are not significant or at a time when the sections do not comprise sufficient number of observations.

The described procedure utilizes the fact that $S_{1k}^*(G)$ and $S_{1l}^*(G)$ are dependent only through $\bar{a}_n$ for $|k - l| > 2G$, which means that they can be considered to be independent for n large. The same holds for $S_{2k}^*(G)$ and $S_{2l}^*(G)$. This fact allows to control the level of such multistage tests.

For example, (10.20) can be this way used to detect not only a beginning of linear trend but also its end. To recognize whether there is a decrease or an increase one need to consider formulas (10.13) through (10.15), and (10.19) through (10.21) without absolute values. The use is clear. In Theorems 10.2.1 and 10.2.2 one only have to exchange the limit $\exp(-2e^{-y})$ for $\exp(-e^{-y})$ in case of maximum. For minimum use correspondence $\min X_i = -\max\{-X_i\}$.

The strategy for the choice of G will be demonstrated by the following example of application of (10.21). It can be helpful at investigating efficiency of some measure, for instance in operational risk control where X_i represent the severity of operational losses for the ith day. Suppose the measure were asserted on the first day. The hypothesis of "positive mass" after asserted measure could be assessed like

$$X_i = \theta + \delta\, I_{\{i > m_1\}} + \Delta\, \frac{i - m_2}{n}\, I_{\{m_2 < i < m_3\}} + \Delta\, \frac{m_3 - m_2}{n}\, I_{\{i \geq m_3\}} + e_i,$$
$$i = 1, \ldots, n, \text{ and } \delta > 0, \Delta < 0 \tag{10.32}$$

One could expect that m_1 is not very far from 1 and m_2. It means that G should be chosen small enough to detect these points. Also, G should be smaller than $(m_3 - m_2)/2$. After we assess the largest plausible value of G in view of practical facts we decide whether the limit approximation for critical values can be good for some of these values. Generally consider $G > \sqrt{n}$ for $M_1(G)$ and $G < \sqrt{n}$ for $M_2(G)$. In (Slabý, 1999) it is also shown that G should be large to gain a large asymptotic power of the test (10.1) against (10.2) via $M_1(G)$. Of course, it has to satisfy condition (10.27) at the same time. However, we rather recommend to use simulations to compute critical values instead of the limit approximations. With a good computer it does not take more than several minutes since for 10000 samples the simulated quantiles already seem to be stable. Selected critical values can be found in Table 10.2.

The last practical question is how to choose the scores. The recommendation is that they should be chosen as in situation when the change times are known. For example, if you test alternative (10.2) and expect that the underlying distribution is close to normal then choose van der Waerden scores for the two-sample problem.

	n	C_1	C_2	$M_1(G_1)$	$M_2(G_2)$	Q	$Q^*(G)$
	30	2.49	2.07	2.46	2.49	7.92	7.85
	50	2.59	2.11	2.66	2.64	8.66	8.87
90%	100	2.73	2.18	2.91	2.82	9.67	10.26
	200	2.78	2.21	3.12	2.96	10.36	11.62
	500	2.90	2.27	3.33	3.13	11.17	13.07
	n	C_1	C_2	$M_1(G_1)$	$M_2(G_2)$	Q	$Q^*(G)$
	30	2.70	2.31	2.69	2.66	9.08	8.96
	50	2.81	2.34	2.90	2.81	10.05	10.09
95%	100	2.98	2.44	3.15	3.01	11.06	11.56
	200	3.02	2.45	3.33	3.15	11.79	12.86
	500	3.14	2.54	3.55	3.31	12.63	14.42
	n	C_1	C_2	$M_1(G_1)$	$M_2(G_2)$	Q	$Q^*(G)$
	30	3.12	2.79	3.19	2.91	11.66	10.97
	50	3.27	2.81	3.32	3.12	12.96	12.45
99%	100	3.40	2.89	3.56	3.36	14.37	14.43
	200	3.49	2.94	3.78	3.50	15.14	16.04
	500	3.65	3.01	3.96	3.70	15.82	17.51

Table 10.2: Simulated critical values (Wilcoxon scores)

10.5 Acknowledgements

This work poses a part of my dissertation and was supported by the grant GAČR-201/97/1163 and CES:J13/98:113200008. I would like to thank Marie Hušková, my dissertation supervisor, who meaningfully routed my work and, at the same time, let my ideas fully grow. Her hints and suggestions were very helpful to me. Also, I am grateful to Jaromír Antoch for his hints concerning simulations and Karel Kupka, who revealed me the world of S-Plus programming art.

Bibliography

Bergman, M. (1992). *Sojourns and Extremes of Stochastic Processes*, Wadworth & Brooks/Cole.

Embrechts, P., Klüppelberg, C. and Mikosch, T. (1997). *Modelling Extremal Events for Insurance and Finance*, Springer-Verlag, Berlin Heidelberg.

Hušková, M. (1996). *Limit Theorems for M-processes via Rank Statistics Processes*, Advances in Combinatorial Methods with Applications to Probability and Statistics.

Hušková, M. (1997). Limit theorems for rank statistics, *Statistics & Probability Letters* **32**: 45–55.

Hušková, M. (1998). Estimators in the location model with gradual changes, *Commentationes Mathematicae Universitatis Carolinae* **39,1**: 147–157.

Jarušková, D. (1998). Testing appearance of linear trend, *Journal of Statistical Planning & Inference* **70**: 263–276.

Leadbetter, M., Lindgren, G. and Rootzén, H. (1983). *Extremes and related properties of random sequences and processes*, Springer-Verlag, Heidelberg.

Slabý, A. (1999). *The method of mosum and its modifications*, Master's thesis, A diploma thesis, The Faculty of Mathematics and Physics, Charles University, in Czech.

11 A stable CAPM in the presence of heavy-tailed distributions

Stefan Huschens and Jeong-Ryeol Kim

11.1 Introduction

One of the main assumptions of CAPM is the normality of returns. A powerful statistical argument for the Gaussian assumption is the Central Limit Theorem (CLT), which states that the sum of a large number of independent and identically distributed (iid) random variables (r.v.'s) from a finite-variance distribution will tend to be normally distributed. Due to the influential works of Mandelbrot (1963), however, the stable non-Gaussian, or rather, α-stable distribution has often been considered to be a more realistic one for asset returns than the normal distribution. This is because asset returns are typically fat–tailed and excessively peaked around zero—phenomena that can be captured by α-stable distributions with $\alpha < 2$. The α-stable distributional assumption is a generalization rather than an alternative to the Gaussian distribution, since the latter is a special case of the former. According to Generalized CLT, the limiting distribution of the sum of a large number of iid r.v.'s must be a stable distribution, see Zolotarev (1986).

For the estimation of the β-coefficient in the CAPM OLS estimation is typically applied, which according to the Gauss-Markov theorem, has the minimum variance among all linear unbiased estimators when the disturbance follows a distribution with finite variance. When the disturbance follows a distribution with infinite variance, however, the OLS estimator is still unbiased but no longer with minimum variance. By relaxing the normality assumption by allowing disturbances to have a symmetric α-stable distribution with infinite variance, Blattberg and Sargent (1971) generalize the OLS estimator (henceforth, referred to as the BS estimator), which we apply for estimating of β-coefficient in CAPM under the α-stable distributional assumption.

In this paper, we study CAPM under the α-stable distributional assumption. We employ the Q_p-statistic, a simple estimator for α in Huschens and Kim (1998), to apply the BS estimator for data with unknown α. We also use a bivariate symmetry test to check the linearity of stable CAPM's of interest. In an empirical application, we apply a stable CAPM for estimating the β-coefficients for German Stock Index (DAX) and its individual asset returns. It turned out that the estimated values of the β-coefficients are different, depending on the distributional assumption.

The rest of this paper is organized as follows: Section 2 presents empirical evidence for fat–tailed distributions in DAX data and gives a short summary of α-stable distributions. In Section 3, we study a stable CAPM and the BS estimator. Empirical results of the bivariate symmetry test and of the β-coefficient from the stable CAPM, compared to the conventional CAPM, are presented in Section 4. Section 5 summarizes the main result of the paper.

11.2   Empirical evidence for the stable Paretian hypothesis

11.2.1   Empirical evidence

It is well-known that in most financial data, there are high volatile and low volatile phases, and that they alternate periodically. Mandelbrot (1963) observed it and reported that "... large changes tend to be followed by large changes of either sign, and small changes tend to be followed by small changes ...", which is called the *Joshep-effect* in the literature. In the analysis and modeling for conditional distribution, this empirical observation is known as volatility clustering. Based on Jensen's inequality, one can easily show that financial data with volatility clustering are excessively peaked around zero in their unconditional distribution. This phenomenon can be explained through the risk-aversion of investors: The bigger a shock is, the more active investors are, so that more information will be arriving in the transaction markets. This alternating volatility is typical for distributions with heavy tails and at the same time many outliers.

Figure 1a shows logarithm of DAX for the period from January 4, 1960 until September 9, 1995. Figure 1b shows returns of the DAX. They clearly show volatility clustering. Figure 1c shows the empirical density of the returns which is leptokurtic and heavy-tailed distributed.

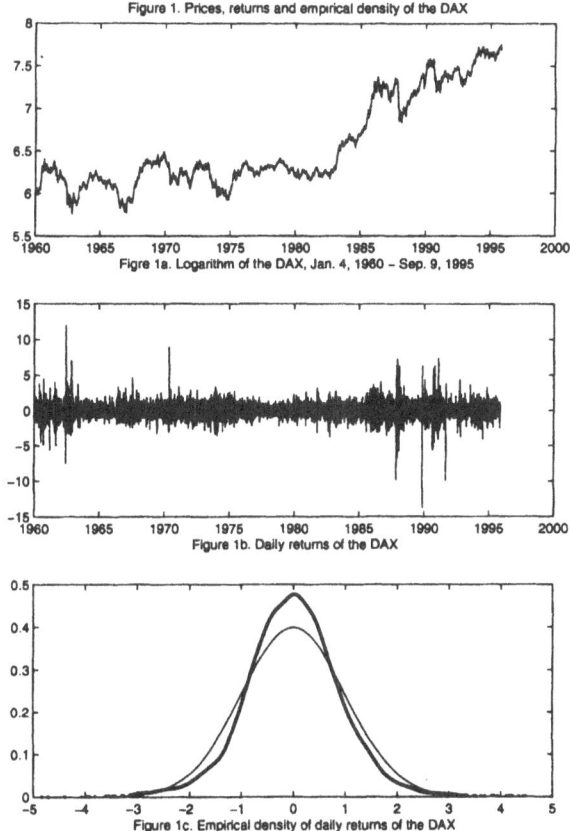

Figure 1. Prices, returns and empirical density of the DAX

Figre 1a. Logarithm of the DAX, Jan. 4, 1960 – Sep. 9, 1995

Figure 1b. Daily returns of the DAX

Figure 1c. Empirical density of daily returns of the DAX

For more empirical evidence of heavy tails in financial data, we refer to Kim (1999a).

11.2.2 Univariate und multivariate α-stable distributions

The shape parameter for the α-stable distribution, α, is called the *stability parameter*. In this paper, we only consider the case for $\alpha > 1$ because of empirical relevance. For $\alpha < 2$, the tails of the distribution are fatter than those of the normal distribution, with tail-thickness increasing as α decreases. If $\alpha < 2$, moments of order α or higher do not exist. A Gaussian r.v. is the special case of an α-stable r.v. when $\alpha = 2$. A stable r.v. with a stability parameter α is said to be α-*stable* For more details on α-stable distributions, see Samorodnitsky and Taqqu (1994). Closed-form expressions of α-stable distributions exist only for $\alpha = 2, 1$

and 0.5. However, the characteristic function of the α-stable distribution can be written as the following (see Samorodnitsky and Taqqu (1994), p. 5):

$$Ee^{iXt} = \exp\left\{-\delta^\alpha|t|^\alpha[1 - ib\,\text{sign}(t)\tan\frac{\pi\alpha}{2}] + i\mu t\right\}, \text{ for } 1 < \alpha \le 2,$$

where μ is the *location parameter*, δ is the *scale parameter*, and $b \in [-1, 1]$ is the *skewness parameter*,[1] which determines the relative size of the two tails of the distribution. If $b = 0$, the distribution is symmetric, denoted as $S_\alpha S$. In this paper we only consider the symmetric case for both univariate and multivariate distributions. This is partly for simplicity and partly for empirical importance. The shape of a stable distribution is, therefore, completely determined by its stability parameter α, when $b = 0$. For $\alpha = 2$ and $b = 0$, we have the normal distribution $N(\mu, 2\delta^2)$.

Figure 2 shows some selected α-stable distributions. The α-stable distributions with $\alpha < 2$ look more like the empirical data in Figure 1c than the normal distribution does. This similarity between α-stable distributions and empirical financial data is an important motivation for applying α-stable distributions.

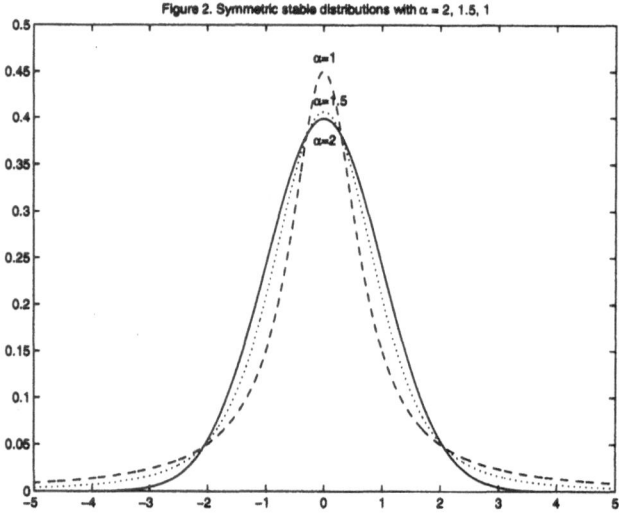

Figure 2. Symmetric stable distributions with $\alpha = 2, 1.5, 1$

---

[1] Usually, β (instead of b) stands for the skewness parameter for the α-stable distribution, but we use b to avoid notational confusion with the β in CAPM.

The characteristic function of an α-stable vector, $\mathbf{X} \in \mathbb{R}^d$, can be written for $1 < \alpha \leq 2$ as the following (for further details, see Samorodnitsky and Taqqu, 1994, p. 73):

$$Ee^{i(\mathbf{X},\mathbf{t})} = \exp\left\{ \int_{S_d} |\cos(\mathbf{t},\mathbf{s})|^\alpha \left(1 - i\ \text{sign}(\mathbf{t},\mathbf{s}) \tan\left(\frac{\pi\alpha}{2}\right)\right) \Gamma(d\mathbf{s}) + i(\mathbf{t},\mu_0) \right\},$$

where Γ is a finite spectral measure of the α-stable random vector on the unit sphere S_d, with $S_d = \{\mathbf{s} : \|\mathbf{s}\| = 1\}$ being the unit sphere in $\mathbb{R}^d$, μ_0 a shift vector, and $\mu_0 \in \mathbb{R}^d$. One important difference between multivariate α-stable distributions and multivariate normal distributions is that multivariate α-stable distributions are in general not completely characterized by a simple covariation matrix.[2] If X_1 and X_2 are iid stable r.v.'s with $\alpha < 2$, the joint distribution will not have circular density contours. In this paper we only consider strictly α-stable vectors, i.e., $\mu_0 = 0$. More exactly, we only consider symmetric, strictly α-stable vectors with $1 < \alpha \leq 2$, i.e., $\mu_0 = 0$ and Γ is a symmetric spectral measure on S_d. We discuss this point later in relation to the linearity of stable CAPM and the bivariate regression curve.

11.3   Stable CAPM and estimation for β-coefficients

The theory of portfolio choice is based on the assumption that investors allocate their wealth across the available assets so as to maximize expected utility. Based on risk aversion of investors, Sharpe (1964) and Lintner (1965) introduce a model of asset returns as a normalized covariance of returns of an asset with returns of a market portfolio. In the frame of the conventional CAPM in which expected utility is a function only of the portfolio's mean and variance, one can formulate the expected return of an asset as follows:

$$E[R_i] = R_f + \beta_i(E[R_m] - R_f) \tag{11.1}$$

with

$$\beta_i = \frac{\text{Cov}[R_i, R_m]}{\text{Var}[R_m]}, \tag{11.2}$$

where R_m is the random return on the market portfolio, and R_f is the return on the risk–free rate of interest. In the conventional CAPM in (11.2), R_i and R_m are typically bivariate normally distributed. The

---

[2] For stable distributions, covariation is analogous to the covariance of the normal distributions.

CAPM implies that the expected return of an asset must be linearly related to the return of the market portfolio.

In order to estimate the β-coefficients, it is necessary to add an assumption concerning the time-series behavior of returns and estimate the model over time. One typically assumes that returns are normal and iid through time. Furthermore, under the assumption of constant covariation structure (otherwise, it would result in a time-varying β), the usual estimator of the asset β is the OLS estimator of the slope coefficient in the excess-return market equation, i.e., the β in the linear regression model with the following time index:

$$y_t = \beta x_t + u_t, \tag{11.3}$$

where u_t is typically assumed as iid with finite variance and expectation zero. The OLS estimator for β is given as

$$\hat{\beta}_{OLS} = \frac{\sum_t x_t y_t}{\sum_t x_t^2} \tag{11.4}$$

and is a best linear unbiased estimator (BLUE). In the context of stable distributions, more general concepts are needed because of infinite variance.

11.3.1 Stable CAPM

Based on empirical evidence, we now turn to α-stable distributed returns, and consequently, to stable CAPM. If asset returns are no longer assumed to be normally distributed, it has consequences on the conventional CAPM, which is based on the existence of second order of moments for the underlying returns.

Fama (1970) considers the conventional CAPM under the special case of multivariate α-stable distributed returns. The return of the ith-asset is generated by the so-called market model

$$R_i = a_i + c_i M + \varepsilon_i, \tag{11.5}$$

where a_i and c_i are asset-specific constants, M is a common market factor, and M and ε_i are mutually independent $S_\alpha S$ r.v's. Under the existence of a riskless asset return R_f and the assumption of risk aversion, the market return of N assets in the market portfolio can be written by

$$R_m = g'R = a_m - c_m M + \varepsilon_m, \tag{11.6}$$

where $g = (g_1, \ldots, g_N)'$ is the proportion of the ith asset in the market portfolio, $a_m = g'a$ with $a = (a_1, \ldots, a_N)$, $c_m = g'c$ with $c = (c_1, \ldots, c_N)$, and $\varepsilon_m = g'\varepsilon$ with $\varepsilon = (\varepsilon_1, \ldots, \varepsilon_N)$. An efficient portfolio minimizes the scale parameter given the mean return. The market portfolio is an efficient portfolio with $(E[R_m] - R_f)/\delta(R_m)$ as slope of the tangent in $(E[R_m], \delta(R_m))$ (see Fama, 1970, p. 49). This implies that

$$_\alpha \beta_i = \frac{1}{\delta(R_m)} \frac{\partial \delta(R_m)}{\partial g_i}. \tag{11.7}$$

is established as an analog to the β-coefficient in (11.2) under the α-stable assumption.

In the bivariate strictly $S_\alpha S$ case with $\alpha > 1$, i.e., $(R_i - E[R_i], R_m - E[R_m])$, Kanter (1972) shows an equivalent expression to (11.7) that

$$_\alpha \beta_i = \frac{E[R_i - E[R_i]|R_m - E[R_m]]}{R_m - E[R_m]} = \frac{\int_0^{2\pi} \sin\theta(\cos\theta)^{<\alpha-1>} d\Gamma(\theta)}{\int_0^{2\pi} |\cos\theta|^\alpha d\Gamma(\theta)},$$

where the quantity $\int_0^{2\pi} \sin\theta(\cos\theta)^{<\alpha-1>} d\Gamma(\theta)$ is called the covariation of two α-stable distributed r.v.'s and $a^{<p>} := |a|^p \text{sign}(a)$. For more details on covariation, see Samorodnitsky and Taqqu (1994), p. 87.

Hardin Jr, Samorodnitsky and Taqqu (1991) examine the determination of the form of regressions under the α-stable distributions and give necessary and sufficient conditions for its linearity, which is implicitly assumed in the formulation of the mean-variance efficient CAPM. One of the results of their analysis is that, as is shown in Kanter (1972), the regression curve is linear if the bivariate α-stable r.v. with $1 < \alpha \le 2$ is symmetric.

Employing the concepts of variation and covariation in Samorodnitsky and Taqqu (1994), Gamrowsky and Rachev (1994) notice a stable CAPM as an analog of the conventional CAPM with the following $_\alpha\beta_i$-coefficient:

$$_\alpha \beta_i = \frac{[R_i, R_m]_\alpha}{[R_m, R_m]_\alpha}, \tag{11.8}$$

which is again an equivalent expression for (11.8).

To sum this up, we can write a stable CAPM as:

$$E[R_i] = R_f + {}_\alpha\beta_i(E[R_m] - R_f), \tag{11.9}$$

where the excess return, $(E[R_i] - R_f, E[R_m] - R_f)$, is bivariate strictly $S_\alpha S$ distributed.

The $_\alpha\beta$ in (11.8), however, does not give a workable formula for estimating the β-coefficient in (11.2) under the α-stable distributional assumption. The BS estimator in Blattberg and Sargent (1971) was firstly employed by Huschens and Kim (1999) to estimate the β-coefficient in the stable CAPM, which is again a stable analog of the normalized covariance for stable laws. In the next subsection we discuss estimation for β-coefficients in the stable CAPM.

11.3.2   Estimation of the β-coefficient in stable CAPM

When $\alpha < 2$, the OLS estimator in (11.4) follows stable distributions with the same stability parameter α as the underlying disturbances-i.e., OLS is very sensitive to outliers. As long as $\alpha > 1$, the OLS will converge to the true parameter value as the sample becomes large, but only at the rate $n^{1-(1/\alpha)}$ rather than the $n^{1/2}$ rate for the Gaussian case. Furthermore, conventional t statistics for estimates of β will no longer have the (usual) Student's t-distribution, but will be concentrated in the vicinity of ± 1, see Logan, Mallows, Rice and Shepp (1973).

Blattberg and Sargent (1971) derive BLUE for the coeficients in linear regressions with deterministic regressor and α-stable distributed disturbances. By minimizing the dispersion parameter under the α-stable distributional assumption, the BS estimator is BLUE for a regression with deterministic regressors and obtained as

$$\beta_{BS} = \frac{\sum_t |x_t|^{1/(\alpha-1)}\text{sign}(x_t)y_t}{\sum_t |x_t|^{\alpha/(\alpha-1)}}, \tag{11.10}$$

which coincides with (11.4) for $\alpha = 2$. The efficiency of the BS estimator, both for the case of deterministic regressor and of stochastic regressors, is demonstrated in Huschens and Kim (1999) by simulation.

The BS estimator assumes that the stability parameter of the underlying data is known. For empirical works, Huschens and Kim (1998) propose a simple quantile statistic for estimating an unknown α. The quantile statistic is a ratio of an estimate of the p-quantile and of the first absolute moment of a distribution truncated on the left at x_{1-p} and on the right at x_p:

$$\hat{Q}_p = \frac{\hat{X}_p}{\sum_{i=1}^n |X_i|\mathbf{I}_{[0,\hat{X}_p]}(|X_i|)/\sum_{i=1}^n \mathbf{I}_{[0,\hat{X}_p]}(|X_i|)}. \tag{11.11}$$

Using the following response surface regression for $p = 0.99$ one can easily check the stability parameter of the data of interest.

$$\hat{\alpha} = -5.60\hat{Q}_{.99}^{-0.5} + 58.45\hat{Q}_{.99}^{-1} - 123.93\hat{Q}_{.99}^{-1.5} + 86.31\hat{Q}_{.99}^{-2},$$

for $1 < \alpha \leq 2$.

11.4 Empirical analysis of bivariate symmetry test

In this section, we present empirical results of the stable CAPM. In the first part of the section, we give a test procedure for bivariate symmetry considered in Kim (1999b). The bivariate symmetry is of interest as a pre-test to ascertain the linearity CAPM, as is discussed in Kanter (1972) and Hardin Jr et al. (1991). In the second part of the section, we present the estimates for β-coefficients in a stable CAPM. For our empirical analysis, we take DAX-30 standard assets for the period from July 20, 1998 till July 18, 1999, in which no changes occured in the composition of the DAX-30.

11.4.1 Test for bivariate symmetry

Using a statistic based on a spherically trimmed mean and a spectral measure for testing general multivariate stable laws in Heathcote, Rachev and Cheng (1991), Kim (1999b) gives the following workable procedure for testing for bivariate α-stable symmetry:

(i) For every observation $[x_{1t}, x_{2t}]$ of $\mathbf{X}_t$, we write the polar coordinates $\rho_t := \sqrt{x_{1t}^2 + x_{2t}^2}$ and inverse tangent $\tilde{\eta}_t := \arctan(x_{1t}/x_{2t})$.

(ii) Let k be a sequence of integers satisfying $1 \leq k \leq n/2$ with n being the sample size, and derive the estimator for the normalized spectral measure

$$\phi_n(\eta) = \frac{1}{k} \sum_{k=1}^{n} \mathrm{I}_{[0,\eta]}(\eta_t) \mathrm{I}_{[\rho_{n-k+1:n}, \infty)}(\rho_t), \quad \eta \in (0, 2\pi], \quad (11.12)$$

where $\mathrm{I}(\cdot)$ is the usual indicator function; and $\rho_{i:n}$ denotes the i-th order statistic. Parameter η_t in (11.12) is defined as

$$\eta_t := \begin{cases} \tilde{\eta}_t, & \text{for } x_{1t}, x_{2t} \geq 0, \\ \pi - \tilde{\eta}_t, & \text{for } x_{1t} < 0, x_{2t} \geq 0, \\ \pi + \tilde{\eta}_t, & \text{for } x_{1t}, x_{2t} < 0, \\ 2\pi - \tilde{\eta}_t, & \text{for } x_{1t} \geq 0, x_{2t} < 0. \end{cases}$$

In practice, one may take the grid $(\eta_1, \cdots, \eta_d)$, $\eta_1 = 2\pi/d$, $\eta_d = 2\pi$, where d is the number of grid points and $2\pi/d$ the step width.

(iii) Under some regularity conditions, one can use the sample supremum of $\phi_n(\eta)$ in the region $0 < \eta \leq \pi$, namely

$$\Phi_n := \sup_{0<\eta\leq\pi} \sqrt{k}\frac{|\phi_n(\eta) - \phi_n(\eta + \pi) + \phi_n(\pi)|}{\sqrt{2\phi_n(\eta)}}, \qquad (11.13)$$

as test statistic.

From the functional limit theorem for $\phi_n(\eta)$, one can easily verify that Φ_n asymptotically follows a standard normal distribution.

Table 11.1 shows the empirical results of testing for bivariate α-stable symmetry. The null hypothesis of bivariate α-stable symmetry cannot be rejected for almost any returns at the 95% significance level. For only 2 of 30 returns (Schering and Thyssen), symmetry cannot be rejected at the 90% significance level. The result indicates a linear relation of stable CAPM of interest in our empirical analysis.

11.4.2   Estimates for the β-coefficient in stable CAPM

Table 11.2 shows the empirical result of estimated β's in the stable CAPM. All estimates of OLS and BS are significant from zero, where the significance of BS estimates is based on the simulated critical values depending on α. For details, see Kim (1999a). For a small α, it is more likely that the difference between two estimates from OLS and BS etimation can be observed. It is noteworthy that the differences from two estimates are rather small, because the differences are based on the loss of efficiency of the OLS under the α-stable distributional assumption. Note that OLS is still unbiased and consistent. The relative efficiency of the OLS compared with BS, measured by the distance between the 5% and 95% quantil, is 0.996, 0.982, 0.938, 0.923, 0.821, 0.710, 0.617, 0.505 and 0.420 for $\alpha = 1.9, 1.8, 1.7, 1.6, 1.5, 1.4, 1.3, 1.2$, and 1.1, respectively. In practical financial management, however, a small difference in the estimated value for βs can have a crucial effect on decisional processes. More than two thirds of estimated α-values lie in the regions which are more than two Monte Carlo standard deviations away from $\alpha = 2$. It is worthwhile noting that the stability parameter estimation needs a very large sample size. A sample size of 250, as our empirical size, is too small to estimate the stability parameter accurately, and hence, the Monte-Carlo standard deviations are large. Additionally, it is not unusual that stability parameter estimates over 2 are to be expected for iid stable samples with α as low as 1.65, as argued in McCulloch (1997).

11.5 Summary

Based on the empirical evidence, we surveyed a possible stable CAPM with $S_\alpha S$-distributed returns in this paper. To estimate the β-coefficients in the stable CAPM, we applied the BS etimator for a linear regression with disturbances following $S_\alpha S$-distributions. Empirical results show some differences between the estimators from the OLS and BS estimations which result from the efficiency loss of the OLS estimator.

Bibliography

Blattberg, R. and Sargent, T. (1971). Regression with non-Gaussian stable disturbances: Some sampling results, *Econometrica* **39**: 501–510.

Fama, E. (1970). Stable models in testable asset pricing, *Journal of Political Economy* **78**: 30–55.

Gamrowsky, B. and Rachev, S. (1994). Stable models in testable asset pricing, *in* G. Anastassiou and S. Rachev (eds), *Approximation, Probability and Related Fields*, Plenum Press, New York.

Hardin Jr, C., Samorodnitsky, G. and Taqqu, M. (1991). Non-linear regression of stable random variables, *The Annals of Applied Probability* **1**: 582–612.

Heathcote, C., Rachev, S. and Cheng, B. (1991). Testing multivariate symmetry, *Journal of Multivariate Analysis* **54**: 91–112.

Huschens, S. and Kim, J.-R. (1998). Measuring risk in value–at–risk in the presence of infinite–variance, *Technical report*, 25/98, Department of Statistics, Technical University of Dresden.

Huschens, S. and Kim, J.-R. (1999). BLUE for β in CAPM with infinite variance, *Technical report*, 26/99, Department of Statistics, Technical University of Dresden.

Kanter, M. (1972). Linear sample spaces and stable processes, *Journal of Functional Analysis* **9**: 441–459.

Kim, J.-R. (1999a). *A Generalization of Capital Market Theory — Modeling and Estimation of Empirically Motivated CAPM*, in preparation.

Kim, J.-R. (1999b). Testing for bivariate symmetry: An empirical application, *Mathematical and Computer Modelling* **29**: 197–201.

Lintner, J. (1965). The valuation of risk assets and the selection of risky investments in stock portfolios and capital budgets, *Review of Economics and Statistics* **47**: 13–37.

Logan, B., Mallows, C., Rice, S. and Shepp, L. (1973). Limit distributions of self-normalized sums, *Annals of Probability* **1**: 788–809.

Mandelbrot, B. (1963). The variation of certain speculative prices, *Journal of Business* **26**: 394–419.

McCulloch, J. (1997). Measuring tail thickness to estimate the stable index α: A critique, *Journal of Business and Economic Statistics* **15**: 74–81.

Samorodnitsky, G. and Taqqu, M. (1994). *Stable Non-Gaussian Random Processes*, Chapman & Hall, New York.

Sharpe, W. (1964). Capital asset prices: A theory of market equilibrium under conditions of risk, *Journal of Finance* **19**: 425–442.

Zolotarev, V. (1986). *One-dimensional Stable Distributions. Translations of Mathematical Monographs Vol. 65*, American Mathematical Society, Providence.

Table 11.1: Results of the test for bivariate symmetry

Test Daten	Φ-Statistic (*p*-value)
Adidas-Salomon	1.11 (0.27)
Allianz	0.71 (0.48)
BASF	1.22 (0.22)
Bayer	1.06 (0.29)
BMW	1.00 (0.32)
Commerzbank	1.41 (0.16)
DaimlerChrysler	1.63 (0.10)
Degussa	1.42 (0.15)
Dresdner Bank	0.50 (0.62)
Deutsche Bank	0.71 (0.48)
Deutsche Telekom	0.80 (0.42)
Henkel	1.49 (0.14)
Hoechst	1.28 (0.20)
B. Vereinsbank	0.71 (0.48)
Karstadt	1.58 (0.11)
Linde	1.18 (0.24)
Lufthansa	1.07 (0.29)
MAN	1.54 (0.12)
Mannesmann	0.75 (0.45)
Metro	1.22 (0.22)
Münchener Rück	0.58 (0.56)
Preussag	1.37 (0.17)
RWE	1.22 (0.22)
SAP	0.82 (0.41)
Schering	1.70 (0.09)
Siemens	0.71 (0.48)
Thyssen	1.67 (0.10)
VEBA	0.51 (0.61)
VIAG	1.58 (0.11)
Volkswagen	0.55 (0.58)

Table 11.2: Estimated β- and $_\alpha\beta$-coefficients

Daten	OLS	$\hat{\alpha}$	BS
Adidas-Salomon	0.6926(0.0884)[a]	2.00(0.04)[b]	0.6926(7.84;2.58)[c]
Allianz	1.1470(0.0518)	1.77(0.11)	1.1734(22.65;2.91)
BASF	0.7436(0.0533)	1.67(0.13)	0.7365(13.81;3.40)
Bayer	0.7065(0.0578)	1.91(0.07)	0.7003(12.12;2.56)
BMW	1.1076(0.0800)	1.78(0.10)	1.1088(13.86;2.88)
Commerzbank	0.9878(0.0522)	1.67(0.13)	0.9883(18.94;3.40)
DaimlerChrysler	1.0919(0.0509)	1.87(0.08)	1.1021(21.65;2.61)
Degussa	0.4949(0.0938)	2.00(0.04)	0.4949(5.27;2.58)
Dresdner Bank	1.2110(0.0770)	1.73(0.11)	1.1682(15.16;3.09)
Deutsche Bank	1.0626(0.0638)	1.89(0.07)	1.0541(16.52;2.58)
Deutsche Telekom	1.1149(0.0792)	2.00(0.04)	1.1149(14.08;2.58)
Henkel	0.8582(0.0831)	1.75(0.11)	0.8528(10.26;3.00)
Hoechst	0.9032(0.0776)	1.99(0.04)	0.9016(11.62;2.58)
B. Vereinsbank	1.0555(0.0968)	2.00(0.04)	1.0555(10.91;2.58)
Karstadt	0.6607(0.0807)	1.92(0.06)	0.6593(8.17;2.58)
Linde	0.6668(0.0765)	1.84(0.09)	0.6718(8.79;2.68)
Lufthansa	1.0188(0.0716)	1.87(0.08)	1.0281(14.37;2.61)
MAN	0.8007(0.0841)	2.00(0.04)	0.8007(9.52;2.58)
Mannesmann	1.1836(0.0787)	1.83(0.09)	1.1833(15.04;2.71)
Metro	0.6424(0.0689)	1.76(0.11)	0.6449(9.36;2.96)
Münchener Rück	1.1333(0.0631)	1.95(0.06)	1.1331(17.97;2.58)
Preussag	0.7160(0.0782)	1.90(0.07)	0.7175(9.17;2.58)
RWE	0.6574(0.0820)	1.70(0.12)	0.6280(7.66;3.24)
SAP	1.2553(0.0995)	1.77(0.11)	1.2618(12.68;2.91)
Schering	0.5802(0.0561)	1.97(0.05)	0.5795(10.32;2.58)
Siemens	0.8895(0.0727)	1.76(0.11)	0.8770(12.06;2.96)
Thyssen	0.8082(0.0730)	1.73(0.11)	0.7830(10.72;3.09)
VEBA	0.7447(0.0703)	1.92(0.06)	0.7473(10.63;2.58)
VIAG	0.7898(0.0704)	1.87(0.08)	0.8003(11.36;2.61)
Volkswagen	1.1874(0.0622)	2.00(0.04)	1.1874(19.08;2.58)

[a]Standard deviations are given in parentheses. [b]Monte-Carlo standard deviations are given in parentheses. [c]The first column in parenthesis is the t-statistic and the second is the corresponding critical value.

12 A Tailored Suit for Risk Management: Hyperbolic Model

Jens Breckling, Ernst Eberlein and Philip Kokic

12.1 Introduction

In recent years the need to quantifying risk has become increasingly important to financial institutions for a number of reasons: the necessity for more efficient controlling due to globalisation and sharply increased trading volumes; management of new financial derivatives and structured products; and enforced legislation setting out the capital requirements for trading activities.

As mentioned in Ridder (1998) "the idea of 'Value at Risk' (*VaR*) reflects the industry's efforts to develop new methods in financial risk management that take into account available knowledge in financial engineering, mathematics and statistics". Three standard methods are currently used to evaluate market risk: historical simulation, which in principle is a bootstrap approach, the variance-covariance approach that is also called 'delta normal method', and Monte Carlo simulation. For an in-depth presentation of these techniques the reader is referred to Jorion (1998) and Dowd (1998).

Risk, however, is multifaceted, and it has been shown elsewhere (e.g. Artzner, Delbaen, Eber and Heath (1997)) that *VaR* alone can be deficient in certain regards. A natural property a risk measure is expected to satisfy is subadditivity: the risk of a portfolio should be smaller than the sum of risks associated with its subportfolios. This can also be expressed as: it should not be possible to reduce the observed risk by dividing a given portfolio into subportfolios. In this sense, *VaR* is not subadditive. For this reason other definitions of risk can and should be used depending on the circumstances.

The entire stochastic uncertainty (risk) that is associated with a particular portfolio for a set time horizon is encapsulated within its P&L distribution $F(x)$ (Kokic, Breckling and Eberlein (1999)). For any profit x the function $F(x)$ gives the probability of obtaining no greater profit than x over the time horizon. Thus the most desirable distribution functions are those which increase most slowly and consequently are depicted below all other curves that represent alternative portfolios.

By reading the appropriate quantile value, *VaR* can be obtained directly from the P&L function. While for certain operations of a financial institution *VaR* is a suitable measure of risk, for other operations risk may have to be defined differently. In the classical capital asset pricing approach (Huang and Litzenberger (1988), p. 98), for example, risk is measured in terms of standard deviation (or volatility) of the portfolio. The advantage of centering the analysis on the P&L distribution is that all risk measures of interest are just specific functions of $F(x)$.

A fully-fledged risk management system should therefore enable the user to define risk his/her own way as a function of $F(x)$. Moreover, rather than focusing on risk alone, it may be warranted to relate it to chance. The most common measure of chance used in financial analysis is the *expected return* from a portfolio over a given time frame, although chance could also be defined as the median return, for example, which is a far more robust measure than the mean. These, like any other meaningful measure of chance, can also be expressed in terms of the P&L distribution function. In the context of portfolio management it is equally important to look at the chance side as well as the risk side of the return distribution. Changing a portfolio in order to alter its risk exposure will typically affect the chances as well. How credit risk can be consistently incorporated into this framework, is outlined in section 6, while section 7 presents an example to demonstrate the merits of using the hyperbolic model to describe the P&L distribution.

12.2   Advantages of the Proposed Risk Management Approach

Aspects of a modern risk methodology can be summarized as follows: recognition of the fact that risk assessment actually amounts to a forecasting problem; no assumption of a symmetric P&L distribution, enabling an adequate account of fundamental and derivative securities within the same portfolio; consistent treatment of market and credit risk; an explicit account of inter-market dependencies, allowing for conditional risk assessment (e.g. what would happen to risk if the US dollar was to rise by 1 per cent); flexibility to define appropriate summary statistics according to the task that is to be performed; confidence bounds on the model that is being fitted and on all derived summary statistics; an efficient assessment of the 'goodness-of-fit' of the under-lying model; suggestion of optimal hedge portfolios on the basis of an 'on-the-fly' specified hedge universe (this, at the same time, defines a neat bridge to the area of portfolio management); decomposition of time series into independent components such as an ordinary, a periodic and an outlier component; and risk assessment in real time.

Current risk methodology hardly satisfies any of these properties. For example, most often they concentrate on just a few statistics, such as *VaR*, and the extensive use of Monte-Carlo methods prevents a risk assessment in real time. However, by explicitly forecasting the entire P&L distribution potentially all of the problems above can be solved in due course.

In the following sections a powerful statistical technique for forecast-ing the P&L distribution is introduced. One of its advantages over more conventional methods is that it no longer depends on symmetry assump-tions. This makes it possible to embed all derivatives such as futures, options, swaps etc. in the same framework and analyse them simulta-neously. By taking all statistical and functional relationships between markets into account, different risk profiles compared with conventional approaches emerge, giving rise to a much more realistic risk assessment and more efficient hedging techniques.

In summary, the proposed approach to risk analysis allows one to meet risk limit requirements, to make informed transaction decisions for hedg-ing purposes, and to perform chance/risk optimisation using the pre-ferred definitions of chance and risk. Position risk can be decomposed into its risk element constituents, credit instruments can be naturally embedded, and amongst all permissible transactions the best possible combination that yields the greatest risk reduction can be determined.

12.3   Mathematical Definition of the P & L Distribution

Before these methods can be explained, some notation needs to be intro-duced. Here the notation developed by Ridder (1998) is closely followed.

Let V_t be the market value of a portfolio at time t. Assume that the portfolio consists of J financial instruments and let ω_j, $j = 1, \ldots, J$, be their corresponding weights, or exposures, in the portfolio. In order to obtain a correct risk analysis of the portfolio in its time t state, these weights are held constant throughout the analysis at their time t values.

The stochastic behaviour of the portfolio is determined by the instrument prices P_{jt}, $j = 1, \ldots, J$, which in turn depend on the stochastic behaviour of various underlying risk factors $\mathbf{R}_t = (R_{t1}, \ldots, R_{tK})'$, where $K > 0$ is the number of risk factors covered by the portfolio. For example, these factors could include the prices of underlyings, exchange rates or interest rates amongst others. The instrument prices and hence the portfolio value can be viewed as functions of these risk factors:

$$V_t(\mathbf{R}_t) = \sum_{j=1}^{J} \omega_j \, P_{jt}(\mathbf{R}_t).$$

Initially it is assumed that the log-returns

$$\mathbf{X}_t = (X_{1t}, \ldots, X_{Kt})' = (\ln(R_{1,t}/R_{1,t-1}), \ldots, \ln(R_{K,t}/R_{K,t-1}))'$$

of the risk factors are statistically independent and identically distributed, although it is possible to weaken this condition. Essentially risk arises in the portfolio through adverse movements of the risk factors over time. This results in a change in the value of the portfolio from one time point to the next as given by the one-period profit function:

$$\pi_t(\mathbf{R}_t, \mathbf{X}_{t+1}) = V_{t+1}(\mathbf{R}_{t+1}) - V_t(\mathbf{R}_t).$$

It is also possible to consider different time horizons other than one, but for simplicity of presentation this restriction has been made here.

The conditional distribution of π_t, given all information up to and including time t, is called the P&L-distribution:

$$F_t(x) = P(\pi_t(\mathbf{R}_t, \mathbf{X}_{t+1}) \leq x \mid \mathbf{R}_t, \mathbf{R}_{t-1}, \ldots).$$

The purpose of risk analysis is to measure the *probability and extent* of unfavourable outcomes; in particular outcomes resulting in losses or negative values of π_t. For example, the *'value at risk'* $VaR(p)$ is the limit which is exceeded by π_t (in the negative direction) with given probability p: $F_t(VaR_t(p)) = p$. In practice the computation of $VaR(p)$ is usually performed for $p = 0.01$ and repeated anew each time period.

The true distribution function F_t is usually unknown and needs to be estimated, i.e. forecasted, in order to obtain an estimate of VaR. A very accurate method of estimation is described in the following section.

12.4   Estimation of the P&L using the Hyperbolic Model

A class of distributions, that is tailor-made to capture the uncertainties associated with a financial risk position as described above, are the *generalized hyperbolic distributions*. It was Barndorff-Nielsen (1977) who introduced these distributions in connection with the so-called 'sand project'. Generalised hyperbolic distributions are defined by way of their corresponding *Lebesgue densities*

$$d_{GH}(x; \lambda, \alpha, \beta, \delta, \mu) = a(\lambda, \alpha, \beta, \delta) \left(\delta^2 + (x - \mu)^2 \right)^{(\lambda - 0.5)/2}$$
$$\times K_{\lambda - 0.5} \left(\alpha \sqrt{\delta^2 + (x - \mu)^2} \right) \exp \left(\beta (x - \mu) \right)$$

where

$$a(\lambda, \alpha, \beta, \delta) = \frac{(\alpha^2 - \beta^2)^{\lambda/2}}{\sqrt{2\pi} \ \alpha^{\lambda - 0.5} \ \delta^\lambda K_\lambda \left(\delta \sqrt{\alpha^2 - \beta^2} \right)}$$

is the integration constant and K_λ denotes the modified Bessel function of the third kind and order λ. There are five parameters that determine the *generalized hyperbolic distributions*: $\alpha > 0$ determines the shape, β with $0 \leq |\beta| < \alpha$ the skewness and $\mu \in \Re$ the location, $\delta > 0$ is a scaling parameter, and finally $\lambda \in \Re$ characterises certain subclasses.

There are two alternative parameterisations: (i) $\zeta = \delta \sqrt{\alpha^2 - \beta^2}$ and $\varrho = \beta / \alpha$, and (ii) $\xi = (1 + \zeta)^{-0.5}$ and $\mathcal{X} = \xi \varrho$. The latter is scale- and location-invariant, which means that the parameters do not change if a random variable X is replaced by $aX + b$.

Various special cases of the generalized hyperbolic distributions are of interest. For $\lambda = 1$ one gets the class of *hyperbolic distributions* with densities

$$d_H(x) = \frac{\sqrt{\alpha^2 - \beta^2}}{2 \ \alpha \ \delta \ K_1 \left(\delta \ \sqrt{\alpha^2 - \beta^2} \right)} \ \exp \left(-\alpha \sqrt{\delta^2 + (x - \mu)^2} + \beta \ (x - \mu) \right).$$

This is the class which was introduced in finance by Eberlein and Keller (1995) (see also Eberlein, Keller and Prause (1998)). In the same context Barndorff-Nielsen (1997) investigated the subclass of *normal inverse*

Gaussian (NIG) distributions that results when $\lambda = -\frac{1}{2}$. They can be characterised by the following density

$$d_{NIG}(x) = \frac{\alpha \, \delta}{\pi} \exp\left(\delta\sqrt{\alpha^2 - \beta^2} + \beta(x - \mu)\right) \frac{K_1\left(\alpha g_\delta(x - \mu)\right)}{g_\delta(x - \mu)}$$

where $g_\delta(x) = \sqrt{\delta^2 + x^2}$.

There is a convenient way to derive the densities given above. Generalised hyperbolic distributions are variance–mean mixtures of normal distributions usually denoted by $N(\mu, \sigma^2)$. Let d_{GIG} be the density of the *generalized inverse Gaussian distribution* with parameters $\lambda, \mathcal{X}$ and ψ, i.e.

$$d_{GIG}(x) = \left(\frac{\psi}{\mathcal{X}}\right)^{\lambda/2} \frac{1}{2\,K_\lambda(\sqrt{\mathcal{X}\psi})} x^{\lambda-1} \exp\left(-\frac{1}{2}\left(\frac{\mathcal{X}}{x} + \psi x\right)\right)$$

for $x > 0$. Then it can be shown that

$$d_{GH}(x) = \int_0^{+\infty} d_{N(\mu+\beta y, y)}(x) \, d_{GIG}(y) \, dy$$

where the parameters of d_{GIG} are λ, $\mathcal{X} = \delta^2$ and $\psi = \alpha^2 - \beta^2$.

Knowing the marginal distribution is already half the way to the specification of a *dynamic model* of the asset price process $(R_t)_{t\geq0}$. Though this setting is more general, R_t could be interpreted as one of the risk factors introduced in the last section. Using the common diffusion type approach based on a Brownian motion as the driving process would allow one to obtain a generalized hyperbolic distribution as the stationary distribution of the process, and thus it would appear only in a vague and asymptotic manner. It would be far more desirable, however, if the *log-returns* of the price process $(R_t)_{t\geq0}$ followed this distribution exactly. How this goal can be achieved, is subsequently explained.

Generalised hyperbolic distributions are known to be infinitely divisible (Barndorff-Nielsen and Halgreen (1977)). They thus generate a convolution semigroup of probability measures $(\nu_t)_{t\geq0}$, which is completely determined by the element ν_1, the generalized hyperbolic distribution given by d_{GH}. On the basis of this semigroup a stochastic process $(X_t)_{t\geq0}$ can be constructed in a natural way, that has stationary independent increments $X_t - X_s$ such that $X_0 = 0$ and the distribution of $X_t - X_{t-1}$ for all t is the given generalized hyperbolic distribution.

This process (X_t) will be called the *generalized hyperbolic Lévy motion*. Contrary to the path-continuous Brownian motion it is, excluding the drift component, a purely discontinuous process which can be seen from a careful inspection of the Lévy-Khintchine representation of the Fourier transform of d_{GH}. Hence, the value level of this process is only changed by *jumps*, with an infinite number of small jumps within any finite time interval. Finally, the asset price process $(R_t)_{t \geq 0}$ can now be defined as

$$R_t = R_0 \ \exp(X_t).$$

It follows that the sequence of log-returns $\log R_t/R_{t-1} = \log R_t - \log R_{t-1}$ corresponds to the increments of the generalized hyperbolic Lévy motion $(X_t)_{t \geq 0}$. What are the consequences for estimating the P&L distribution function? To simplify matters, consider a portfolio consisting of just one instrument with a price process given by $(R_t)_{t \geq 0}$. Further assume that the data set consists entirely of *discrete observations*, e.g. daily closing prices $R_0, R_1, R_2, \ldots, R_n$. By setting

$$X_t = \log R_t - \log R_{t-1}$$

the log-returns $X_1, \ldots, X_n$ are derived, which are then used to fit the corresponding empirical distribution function. The generalized hyperbolic parameters can be efficiently estimated via maximum-likelihood, with the resulting generalized hyperbolic distribution determining the return process $(X_t)_{t \geq 0}$. In case of K instruments in the portfolio one could proceed in the same way by considering K-vectors $\mathbf{R}_t = (R_{t1}, \ldots, R_{tK})'$ making up the price process.

Also note that the price process above can be described by a stochastic differential equation. Using Itô's formula for processes containing jumps, it can easily be verified that R_t, as given above, is obtained as the solution of the equation

$$dR_t = R_{t-} \ (dX_t + e^{\Delta X_t} - 1 - \Delta X_t),$$

where $\Delta X_t = X_t - X_{t-}$ denotes the jump at time t.

12.5   How well does the Approach Conform with Reality

Having fitted a distribution, back testing can be used to assess how well the model predicts historical information, when the true outcomes of profit and loss are known exactly. In the '6. KWG-Novelle, Grundsatz 1' requirements a fairly restrictive form of back testing is proposed, where values greater than $VaR(p = 0.01)$ predicted by the model must not be exceeded more than a predefined number of times over a one-year time period. However, for a portfolio manager to base his decisions on (distributional) forecasts, the model must perform well under far more general definitions of back testing. It is of significant importance that the *model predicts profitable outcomes accurately as well as the non-profitable outcomes* that are specifically addressed by the official requirements.

For most models currently implemented it is impossible to perform an anywhere near efficient back testing, because most often only a single quantile is being forecasted rather than the entire distribution, and if distributions are forecasted, then restrictive and unwarranted assumptions are made about its shape. Furthermore, when the predicted value differs from the outcome, for example, it is unclear how significant that difference actually is.

An advantage of forecasting the entire distribution is that one can correctly compare the forecast with the outcome using standard statistical tests, such as the following. Over a given time frame where real outcomes are known let π_t denote the actual outcome of profit/loss at time $t + 1$, and let $\widehat{p}_t = \widehat{F}_t(\pi_t)$ be the percentile corresponding to this value on the basis of the forecasted P&L distribution $\widehat{F}_t$. If the model is correct, then the $\widehat{p}_t$ values should be uniformly distributed on the unit interval $[0, 1]$. Various powerful techniques exist for testing whether a sample is uniformly distributed. In fact, the '6. KWG-Novelle, Grundsatz 1' requirements correspond to assessing whether the proportion of $\widehat{p}_t$ values below 0.01 is close to 1 per cent, which is clearly much weaker than testing the whole $\widehat{p}_t$ distribution.

12.6   Extension to Credit Risk

The inclusion of credit risk in a risk management system has become an increasingly important issue for financial institutions in recent time. However, because of the relatively rare nature of loan defaults, the associated P&L distributions are highly skewed. For this reason the framework of non-normal modelling presented above is highly relevant to the situation of modelling credit risk. In fact it can be shown that the hyperbolic model is highly suited to this situation as its shape readily adapts to highly skewed and heavy-tailed distributions.

In the approach presented above it is assumed that for each instrument, j, there exists a (complete) price series $(P_{jt})_{t=0,1,...}$ without missing values. For market risk this is largely true, and where holes in the data do occur it is possible to fill in the missing gaps using reliable pricing formulae, or by imputation methodology. On the other hand, for credit risk this is generally not the case. Typically the (market) price of a loan is known at the time it is issued, but thereafter it is not and a price must be imputed by some means. Regarding a loan as a defaultable bond, this can be done using the yield-to-price relationship:

$$P_{jt} = \sum_k \frac{c_j(\tau_{jk})}{(1 + \gamma_j(\tau_{jk}))^{\tau_{jk}}},$$

where $c_j(\tau_{jk})$ denotes the cash flow for loan j at forward time τ_{jk} and $\gamma_j(\tau)$ the interest rate of a zero-coupon version of loan j maturing at time τ. The function γ_j can be regarded as a risk factor. The fair price of the loan at the time the loan is taken out is the net present value of a risk-free investment of the same amount, plus a premium to compensate for the probability of default during the loan period, plus a risk premium as reflected in the chance/risk diagram commonly used in portfolio theory.

Typically the yield curve for a loan is not known, which makes continuous pricing difficult. The way that is often employed is that borrowers are categorised by some means and the yield curve for each category is determined on the basis of the yield curve for risk-free investments, on default and recovery rates in the case of default as well as on an (excess chance) amount for the extra risk, and on an assumed stochastic structure for the transition of a loan from one category to another.

To determine a credit portfolio's P&L distribution there are at least two approaches that may be used: the simulation approach and the imputation approach. For a number of fundamental reasons it turns out that the imputation approach is superior to the simulation approach, see Kokic et al. (1999) for details. In particular, using the imputation approach it is possible to determine the P&L distribution of a portfolio made up of a mixture of both credit and market instruments.

12.7   Application

As an illustration of the techniques described in this paper a comparison is made between the variance/covariance and historical simulation models, which are commonly used in most commercially available risk management software, and the hyperbolic model; in particular the *NIG* model which results when $\lambda = -0.5$.

In this example a portfolio will be analysed that has price behaviour identical to the German *DAX* index. Since the *DAX* is a weighted average of 30 liquid share prices, the variance/covariance model might be suspected to work well.

However, if one computes the log-returns of the *DAX* and fits the various distributions to the data, it can readily be seen that there is significant departure from normality, which is implicitly assumed when using the variance-covariance model. Figure 1 shows three distributions fitted to the log-returns of daily *DAX* data between 09.05.1996 and 25.03.1999 (i.e. 750 observations in total). In this Figure the empirical distribution is obtained using a kernel smoother with a fixed bandwidth equal to one fifth of the inter-quartile range. Both, the normal and *NIG* distribution are fitted by maximum likelihood methods.

For small values the *NIG* distribution yields a considerably better fit. In particular, it is able to readily adapt to the left skewness of the data, whereas the normal distribution is not. This feature especially is likely to be crucial for an accurate computation of *VaR* and other 'downside' risk measures. The *NIG* distribution also fits markedly better towards the centre of the data. That is, there is a much larger proportion of small value changes than suggested by the normal distribution.

As a consequence one would expect the back test performance of the hyperbolic model to be better than for the variance/covariance model. To examine this issue in more detail, back testing according to the methodology outlined in section 5 was performed. To be more specific, for any time point t in the back test period, a P&L distribution was estimated from the preceding 750 days of data and, based on the outcome at time t, a $\widehat{p}_t$ value was computed. This process was repeated for each time point t between 16.11.1994 and 25.03.1999, a total of 1134 evaluations.

Three models were tested: the variance/covariance model; the historical simulation model; and finally a 'stabilized' form of the hyperbolic model based on the *NIG* distribution referred to above. The result of the back testing leads to three $\widehat{p}_t$ time series, corresponding to the models fitted.

Figure 12.1: *Estimated density functions for the log-returns of daily DAX data between 09.05.1996 and 25.03.1999 (i.e. 750 observations)*

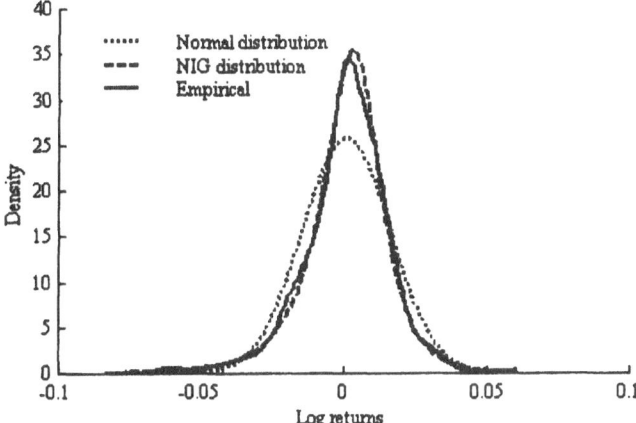

According to the criteria developed in section 5, the model that yields the most uniformly distributed $\hat{p}$ values best captures the real market behaviour.

To get an impression of how uniformly distributed the $\hat{p}$ values are, kernel smoothed densities for each of the three models are presented in Figure 2. The poorest result is obtained for the variance/covariance model, with the corresponding density showing a systematic departure from uniformity across the entire range of p values. As indicated above, a much greater frequency of occurrences can be observed in both tails and in the centre of the P&L distribution than predicted by the variance/covariance model. This poor performance suggests that the use of this model would almost certainly result in severe underestimation of risk.

There appears to be some improvement in the fit with the historical simulation model. In particular, the under-estimation of central events in the P&L distribution that was observed for the variance/covariance model has largely disappeared. However, the fit in the tails is still poor. Although the reason does not seem obvious, most of this deficiency can be attributed to fluctuations in the volatility of the *DAX* time series. The stabilized hyperbolic model is able to take these fluctuations into account and hence captures the distribution of extreme events accurately.

Figure 12.2: *Back test density estimates based on daily DAX data performed from 16.11.1994 to 25.03.1999 (i.e. 1134 evaluations)*

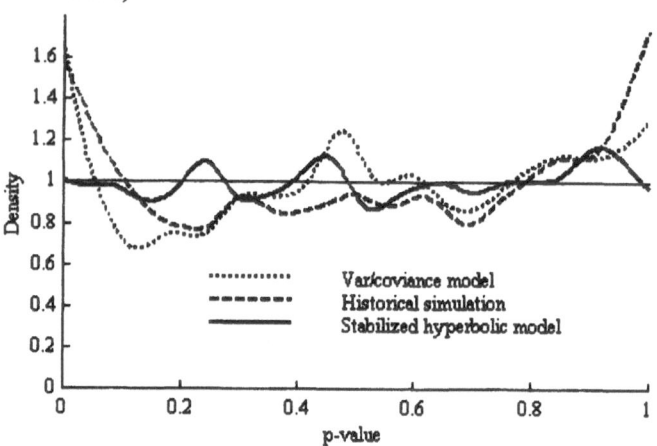

As illustrated in Figure 2, this model yields a superb fit over the entire range of p values.

This is also evident from the results of the four different tests for goodness-of-fit of the various models presented in Table 1. The two values presented are the value of the test statistic itself and the probability value of the test, with small probability values, e.g. less than 0.05, indicating that the model is not forecasting the P&L distribution well. Each test examines a different aspect of the goodness-of-fit of the model. The conventional test is equivalent to testing whether or not the frequency of $\hat{p}_t$ values less than 0.01 is close to 1 per cent, and thus is similar to the test proposed in the '6. KWG-Novelle, Grundsatz 1' requirements. By contrast, both the $\mathcal{X}^2$ and Kolmogorov tests assess the 'uniformity' of the distribution of $\hat{p}_t$ values. That is, the entire range of $\hat{p}_t$ values from zero to one, and not just the $\hat{p}$ values less than 0.01 are examined. The fourth test, finally, is to check for existence of residual autocorrelation in the time series of p values, which would also be evidence of poor model fit.

The results in Table 1.1 confirm the conclusions drawn from Figure 2: there is evidence that the historic simulation yields a better fit than the variance/covariance model. However, comparing these models with the stabilized hyperbolic model, the latter describes the actual data far better and thus provides a much better assessment of the frequency *and*

Table 12.1: *Goodness-of-fit test statistics / probability values for the various models using daily* DAX *data. Back testing performed from 16.11.94 to 25.03.1999 (i.e. 1134 evaluations)*

Test	Variance–covariance	Historical simulation	Stabilized hyperbolic
Conventional	0.034/0.000	0.025/0.000	0.010/0.844
χ^2	202/0.000	199/0.000	82/0.899
Kolmogorov	0.046/0.018	0.058/0.001	0.016/0.792
Serial corr.	145/0.002	174/0.000	73/0.978

the severity of 'bad' events. In this case all tests are passed comfortably.

Bibliography

Artzner, P., Delbaen, F., Eber, J.-M. and Heath, D. (1997). Thinking coherently, *RISK* **10**(11): 68–71.

Barndorff-Nielsen, O. (1977). Exponentially decreasing distributions for the logarithm of particle size, *Proceedings of the Royal Society London A* (353): 401–419.

Barndorff-Nielsen, O. (1997). Processes of normal inverse gaussian type, *Finance & Stochastics* (2): 41–68.

Barndorff-Nielsen, O. and Halgreen, O. (1977). Infinite divisibility of the hyperbolic and generalized inverse gaussian distributions, *Zeitschrift für Wahrscheinlichkeitstheorie und verwandte Gebiete* (38): 309–312.

Dowd, K. (1998). *Beyond Value At Risk: The New Science of Risk Management*, Wiley&Son.

Eberlein, E. and Keller, U. (1995). Hyperbolic distributions in finance, *Bernoulli* **1**: 281–299.

Eberlein, E., Keller, U. and Prause, K. (1998). New insights into smile, mispricing and value at risk: the hyperbolic model, *Journal of Business* (71): 371–405.

Huang, C. and Litzenberger, R. (1988). *Foundations for financial economics*, North-Holland, New York.

Jorion, P. (1998). *Value at Risk: The New Benchmark for Controlling Derivatives Risk*, McGraw-Hill.

Kokic, P., Breckling, J. and Eberlein, E. (1999). A new framework for the evaluation of market and credit risk. In press.

Ridder, T. (1998). Basics of statistical var-estimation, *in* G. Bol, G. Nakhaeizadeh and K. Vollmer (eds), *Risk measurement, econometrics and neural networks*, Physica-Verlag, Heidelberg, New York.

13 Computational Resources for Extremes

Torsten Kleinow and Michael Thomas

13.1 Introduction

In extreme value analysis one is interested in parametric models for the distribution of maxima and exceedances. Suitable models are obtained by using limiting distributions. In the following lines, we cite some basic results from extreme value theory. The reader is referred to (Embrechts, Klüppelberg and Mikosch, 1997) and (Resnick, 1987) for a theoretical and to (Reiss and Thomas, 1997) for an applied introduction. A more detailed review is given in (McNeil, 1997).

A classical result for the distribution of maxima was given by Fisher and Tippett (1928). Assume that $X, X_1, X_2, \ldots$ are i.i.d. with common distribution function F. If for suitable constants a_n and b_n the standardized distribution of the maximum

$$P\left\{\frac{\max\{X_1, \ldots, X_n\} - b_n}{a_n} \leq x\right\} = F^n(a_n x + b_n)$$

converges to a continuous limiting distribution function G, then G is equal to one of the following types of extreme value (EV) distribution functions.

(i)   Gumbel (EV0) $G_0(x) = \exp(-e^{-x})$, $x \in \mathbb{R}$,
(ii) Fréchet (EV1) $G_{1,\alpha}(x) = \exp(-x^{-\alpha})$, $x \geq 0, \alpha > 0$,
(iii) Weibull (EV2) $G_{2,\alpha}(x) = \exp(-(-x)^{-\alpha})$, $x \leq 0, \alpha < 0$.

By employing the reparametrization $\gamma = 1/\alpha$, these models can be unified using the von Mises parametrization

$$G_\gamma(x) = \begin{cases} \exp(-(1 + \gamma x)^{-1/\gamma}), & 1 + \gamma x > 0, \gamma \neq 0, \\ \exp(-e^{-x}), & x \in \mathbb{R}, \gamma = 0. \end{cases}$$

One says that the distribution function F belongs to the domain of attraction of the extreme value distribution G, in short $F \in \mathcal{D}(G)$. The Gnedenko-De Haan theorem as well as the von Mises conditions provide sufficient conditions for $F \in \mathcal{D}(G)$ (see, e.g., (Falk, Hüsler and Reiss, 1994) for details). Moreover, the assumption of independence can be weakened (see, e.g., (Leadbetter and Nandagopalan, 1989)).

One may also consider the distribution function $F^{[t]} := P(X < \cdot | X > t)$ of exceedances above a threshold t, where F lies in the domain of attraction of the extreme value distribution G_γ. Balkema and de Haan (1974) as well as Pickands (1975) showed that for suitable a_u and b_u the truncated distribution $F^{[u]}(b_u + a_u x)$ converges to a generalized Pareto (GP) distribution W_γ as $u \to \omega(F) := \sup\{x : F(x) < 1\}$, with

$$
W_\gamma(x) = \begin{cases}
1 - (1 + \gamma x)^{-1/\gamma} & x > 0, \gamma > 0 \\
& 0 < x < -1/\gamma, \gamma < 0 \\
1 - e^{-x} & x \geq 0, \gamma = 0.
\end{cases}
$$

Again, by using the parametrization $\alpha = 1/\gamma$, one obtains the three submodels

(i)   Exponential (GP0) $W_0(x) = 1 - e^{-x}$, $x \geq 0$,
(ii)   Pareto (GP1) $W_{1,\alpha}(x) = 1 - x^{-\alpha}$, $x \geq 1, \alpha > 0$,
(iii   Beta (GP2) $W_{2,\alpha}(x) = 1 - (-x)^{-\alpha}$, $-1 \leq x \leq 0, \alpha < 0$.

These limit theorems suggest parametric distributions for data which are block maxima or exceedances above a threshold t. In the next section, we describe a computational approach for fitting these distributions to data.

13.2   Computational Resources

A similar extreme value module is implemented in the two software packages XploRe and Xtremes. We give a short introduction to the systems and provide an overview of the extreme value methods that are implemented.

13.2.1   XploRe

XploRe is an interactive statistical computing environment. It provides an integrated programming language with a large library of predefined functions and interactive tools for graphical analysis. A program written in the XploRe language is called *quantlet*. These quantlets are collected in libraries. The interactive tools include displays, with one or more plots, and low level GUI elements for user interaction during quantlet execution. To use XploRe without writing quantlets, it is possible to execute simple instructions on the command line, such as reading data, loading libraries or applying quantlets from a library to data.

There are two fundamental versions of XploRe. The first is a standalone statistical software available on several computer platforms, while the second one is a client/server system (www.xplore-stat.de). As described in section 13.3.1, the client/server architecture has many advantages. However, due to the early state of development, the XploRe client does not yet provide the same functionality as the standalone application.

13.2.2 Xtremes

The MS-Windows based statistical software Xtremes offers a menu-driven environment for data analysis. Besides the usual visualization options, there are parametric estimation procedures for Gaussian, extreme value and generalized Pareto distributions. Special menus offer applications of extreme value analysis to problems arising in actuarial and financial mathematics as well as hydrology. See www.xtremes.de for more informations.

13.2.3 Extreme Value Analysis with XploRe and Xtremes

Estimators for GP and EV

Various estimators for extreme value and generalized Pareto distributions are implemented. We list the estimators available for GP distributions:

- Hill estimator, which is a m.l. estimator for the Pareto (GP1) submodel,

- m.l. estimator for the full GP model,

- Pickands estimator (see (Pickands, 1975)),

- Drees-Pickands estimator, which uses a convex combination of Pickands estimators (see (Drees, 1995)),

- moment estimator (see (Dekkers, Einmahl and de Haan, 1989)).

Two estimators for the EV distributions are provided:

- M.l. estimator for the full EV model,

- linear combination of ratio of spacings estimator, a construction similar to that of the Pickands estimator.

More details on the estimators are given in the cited literature as well as in Reiss and Thomas (1997) and Reiss and Thomas (1999). While the fitting of an extreme value distribution is straight forward, a generalized Pareto distribution is fitted in two steps.

1. Select a threshold t and fit a GP distribution $W_{\gamma,t,\sigma}$ to the exceedances above t, where γ is the shape parameter and t and σ are location and scale parameter.

2. Transform the distribution to $W_{\hat{\gamma},\hat{\mu},\hat{\sigma}}$ which fits to the tail of the original data. The transformation is determined by the conditions $W_{\hat{\gamma},\hat{\mu},\hat{\sigma}}^{[t]} = W_{\gamma,t,\sigma}$ and $W_{\hat{\gamma},\hat{\mu},\hat{\sigma}}^{[t]}(t) = (n-k)/n$, where n is the sample size and k the number of exceedances above t. One obtains $\hat{\gamma} = \gamma$, $\hat{\sigma} = \sigma(k/n)^{\gamma}$ and $\hat{\mu} = t - (\sigma - \hat{\sigma})/\gamma$ as estimates of the tail fit. The latter values are displayed by the software.

In our implementation, we fix the number of upper extremes and use the threshold $t = x_{n-k+1:n}$.

Choosing a Threshold

The selection of an optimal threshold is still an unsolved problem. We employ a visual approach that plots the estimated shape parameter against the number of extremes. Within such a plot, one often recognizes a range where the estimates are stable. A typical diagram of estimates is shown in section 13.2.3.

Checking the Quality of a Fit

A basic idea of our implementation is to provide the ability to check a parametric modeling by means of nonparametric procedures. The software supports QQ-plots and a comparison of parametric and empiric versions of densities, distribution and quantile functions. An important tool for assessing the adequacy of a GP fitting is the mean excess function. It is given by

$$e_F(t) := E(X - t | X > t),$$

where X is a random variable with distribution function F. For a generalized Pareto distribution W_γ, the mean excess function is

$$e_{W_\gamma}(t) = \frac{1 + \gamma t}{1 - \gamma}.$$

We can therefore check if a GP tail is plausible by means of the sample mean excess function. Moreover, by comparing sample and parametric mean excess functions fitted by an estimator, a visual check of an estimation and a choice between different estimators becomes possible. The following section 13.2.3 demonstrates this approach.

Example Analysis of a Data Set

To exemplify the computational approach, we analyze a data set with the daily (negative) returns of the Yen related to the U.S. Dollar from Dec. 78 to Jan. 91. Figure 13.1 (left) shows a scatterplot of the 4444 returns. A fat tail of the distribution is clearly visible. In the following, we fit a generalized Pareto distribution to the tail of the returns by using the moment estimator. To find a suitable threshold, a diagram of the estimates is plotted in Figure 13.1 (right). For $50 \leq k \leq 200$ the estimates are quite stable.

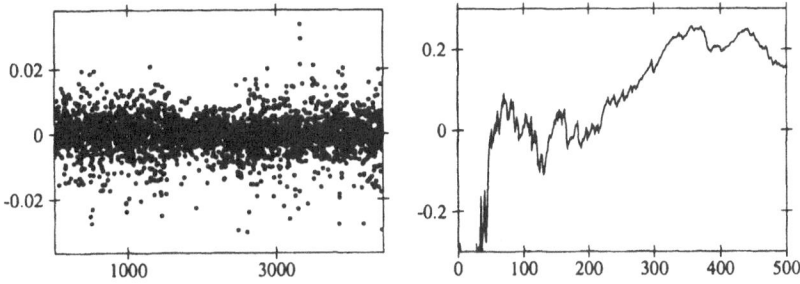

Figure 13.1:   Daily returns of Yen/U.S. Dollar from Dec. 1978 to Jan. 1991 (left) and diagram of estimated shape parameters (right).

We select $k = 160$ extremes, yielding a threshold $t = 0.00966$ and plot a kernel density estimate (solid) as well as the parametric density fitted by the moment estimator (dotted) and the Hill estimator (dashed) for that number of extremes. The resulting picture is shown in Figure 13.2 (left).

Although the parameters estimated by the moment estimator seem to fit the kernel density slightly better, it is not easy to justify a parametric

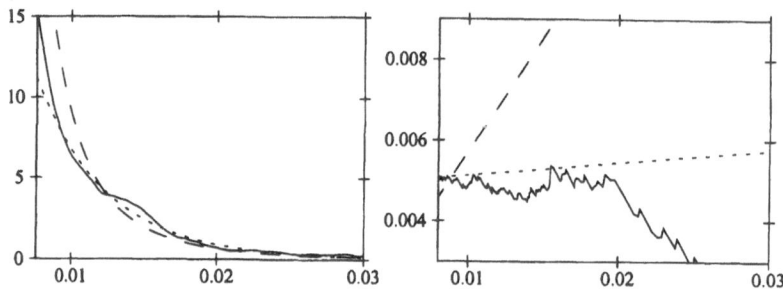

Figure 13.2:   Densities (left) and mean excess functions (right) fitted by moment estimator (dotted) and Hill estimator (dashed).

model from the plot of the densities. We therefore also plot the mean excess functions. The right hand picture in Figure 13.2 shows the empirical mean excess function and the parametric versions, based on the same estimates. While the one fitted by the moment estimator (dotted) is close to the empiric version (solid), the one fitted by the Hill estimator (dashed) shows a strong deviation. This indicates that the parameters obtained by the moment estimator may be more appropriate.

13.2.4   Differences between XploRe and Xtremes

The XploRe system provides the user with an immediate language. Typical features of such a language (according to Huber (1994)) are the omission of declarations and the ability to implement macros using the same constructs as in an immediate analysis.

Xtremes implements a menu interface for interactions and a compiled language for user written routines, whereby the user is required to declare all objects used within a program. That approach results in longer and more complex programs which are typically less flexible than interpreted ones with runtime type checking. However, a higher execution speed can be achieved as syntactic and semantic checks are performed at compile time.

13.3   Client/Server Architectures

Client/server architectures are becoming increasingly important in statistical computing. We discuss two of their advantages which are employed in XploRe and Xtremes: the separation of computational part and user interface and the provision of servers for special, user-written clients.

13.3.1 Client/Server Architecture of XploRe

The client/server version of the XploRe software package consists of three parts. The XploRe server is a batch program which provides several methods for statistical computing. The client is a GUI written in Java providing an interface for the user to interact with the server. Between these two resides a middleware program which manages the communication between client and server. Figure 13.3 shows the structure of this architecture.

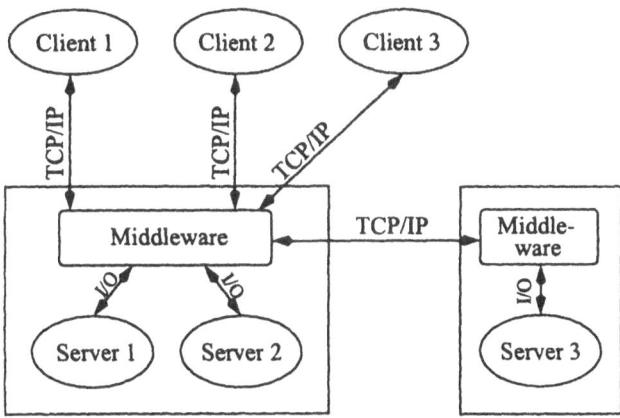

Figure 13.3: The Client/Server architecture of XploRe

Details of the Architecture

The main task of the XploRe server is to provide a statistical programming language and a variety of numerical methods for statistical analysis. To ensure high flexibility, it is possible to add methods (shared libraries, dynamically linked libraries) to the server dynamically. The xtremes library uses this mechanism. The server executes quantlets (programs written in the XploRe language) and writes the output to the standard output stream. Graphical output is encoded in a special protocol which is interpreted by the client.

The client provides the user with a GUI that lets him write and execute programs on a server, show numerical results and display the graphical output of an analysis. The platform independent client runs on every machine where a Java runtime environment is available. The server is written in C and C++, providing the numerical power needed for fast statistical computations.

The central part of this software package is the middleware. It is written in Java and resides on the same host as the server does. Its most important task is the management of the communication between server and client.

Advantages of the Architecture

One of the main advantages of the client/server architecture that is implemented within XploRe is the separation of the computational part and the user interface. It enables the user to use one or more servers without requiring high computational power on the host where the client is running. Instead, he has remote access to statistical methods as well as to computational resources.

In earlier versions of XploRe, the client/server communication has been managed by the server itself. The advantage of the separation of the managing part and the calculation part is a higher stability of the system as well as the opportunity to use different servers with one middleware. These servers could be Gauss, shazam or any other batch program.

Plans for Future Developments

In the future, the middleware should act as a distribution server; i.e., when an old client logs into the middleware, the user is offered an automatic update of the client. The old client downloads a new version from the middleware and installs it on the client host without user interaction. Another task of the middleware will be load average. This means when a middleware is contacted by a client it asks other middleware programs for the load of the hosts where they are running. The requests will then be sent to the host with the smallest load.

Due to the separation of the user interface and the computational part, different clients can be developed. In addition to the usual Java client, a prototype of an MS-Excel add-on exists. Besides clients for special environments (Java/Web, Excel), one could also think of clients for special purposes like finance or time series analysis. A Java-API (Application Program Interface) for the development of clients will be made available in future releases.

13.3.2 Xtremes CORBA Server

CORBA (see (OMG, 1995) or www.omg.org) is a platform and programming language independent standard for distributed, object oriented software systems. It encapsulates all network specific operations. Invoking methods on a remote server object is done by calling methods of a local proxy object. An *interface definition language* (IDL) describes the methods offered by the server objects.

Xtremes implements a CORBA-compliant server which exports statistical procedures (such as estimators or data generation routines). We list an excerpt of the interface definition.

```
enum TPortType {  PORT_INT, PORT_REAL, ... };
typedef sequence<double> doubleseq;

interface TNode {
    string Name ();
    TNode Clone ();

    long GetNumberOfInports ();
    TPortType GetInportType (in long Nr);
    long GetNumberOfOutports ();
    TPortType GetOutportType (in long Nr);

    void SetInport (in long Nr, in any x);
    void Perform ();
    any GetOutport (in long Nr);
};
```

The server objects (called nodes) are self-describing. Besides a name, they return the number and types of parameters and results. After setting the parameters with **SetInport**, the **Perform** method is invoked, and results are fetched by calling **GetOutport**.

The homogeneous structure of the objects facilitates their creation by means of a factory (Gamma, Helm, Johnson and Vlissides, 1995). On startup, the Xtremes server creates a factory object and publishes its object reference.

13.4   Conclusion

We have described two software systems that offer statistical methods for extreme value analysis in a distributed environment. Both systems allow the user to invoke statistical operations from a remote client; yet, different approaches are taken. Future effort should be invested in the specification of a general interface allowing the interoperation of different statistical software packages.

Bibliography

Balkema, A. and de Haan, L. (1974), 'Residual life time at great age.', *Ann. Probab.* **2**, 792–804.

Dekkers, A., Einmahl, J. and de Haan, L. (1989), 'A moment estimator for the index of an extreme-value distribution', *Ann. Stat.* **17**, 1833–1855.

Drees, H. (1995), 'Refined pickands estimators of the extreme value index', *Ann. Stat.* **23**, 2059–2080.

Embrechts, P., Klüppelberg, C. and Mikosch, T. (1997), *Modelling Extremal Events*, Springer.

Falk, M., Hüsler, J. and Reiss, R.-D. (1994), *Laws of Small Numbers: Extremes and Rare Events*, DMV-Seminar, Birkhäuser, Basel.

Fisher, R. and Tippett, L. (1928), 'Limiting forms of the frequency distribution of the largest and smallest member of a sample', *Proc. Camb. Phil. Soc.* **24**, 180–190.

Gamma, E., Helm, R., Johnson, R. and Vlissides, J. (1995), *Design Patterns*, Addison-Wesley, Reading, Massachusetts.

OMG (1995), The common object request broker: Architecture and specification, Technical report, Object Management Group.

Huber, P. (1994), Languages for statistics and data analysis, *in* P. Dirschedl and R. Ostermann, eds, 'Computational Statistics', Physica, Heidelberg.

Leadbetter, M. and Nandagopalan, S. (1989), On exceedance point processes for stationary sequences under mild oscillation restrictions, *in* J. Hüsler and R.-D. Reiss, eds, 'Extreme Value Theory', number 51 *in* 'Lect. Notes in Statistics', Springer, New York.

McNeil, A. (1997), 'Estimating the tails of loss severity distributions using extreme value theory', *ASTIN Bulletin* **27**, 117–137.

Pickands, J. (1975), 'Statistical inference using extreme order statistics', *Ann. Stat.* **3**, 119–131.

Reiss, R.-D. and Thomas, M. (1997), *Statistical Analysis of Extreme Values*, Birkhäuser, Basel.

Reiss, R.-D. and Thomas, M. (1999), Extreme value analysis, *in* 'XploRe -The Statistical Environment', Springer, New York.

Resnick, S. (1987), *Extreme Values, Regular Variation, and Point Processes*, Springer, New York.

14 Confidence intervals for a tail index estimator

Sergei Y. Novak

14.1 Confidence intervals for a tail index estimator

Financial data (log-returns of exchange rates, stock indices, share prices) are often modeled by heavy–tailed distributions, i.e., distributions which admit the representation

$$\mathbf{P}(X > x) = L(x)x^{-1/a} \qquad (a > 0), \tag{14.1}$$

where the function L slowly varies: $\lim_{x \to \infty} L(xt)/L(x) = 1$ $(\forall t > 0)$. The number $1/a$ is called the *tail index* of the distribution (14.1). The problem of estimating the tail index has also important applications in insurance, network modelling, meteorology, etc.; it attracted significant interest of investigators (see Csörgö and Viharos (1998), Embrechts, Klüppelberg and Mikosch (1997), Novak (1996), Resnick (1997) and references therein).

The popular estimator

$$a_n^H \equiv a_n^H(k_n) = k_n^{-1} \sum_{i=1}^{k_n} \ln\left(X_{(i)}/X_{(k_n+1)}\right), \tag{14.2}$$

where $X_{(n)} \le \dots \le X_{(1)}$ are the order statistics, was introduced by Hill (1975). A number of other estimators can be found in Csörgö and Viharos (1998), de Haan and Peng (1998), Resnick (1997).

Goldie and Smith (1987) introduced the *ratio estimator*

$$a_n \equiv a_n(x_n) = \sum_{i}^{n} Y_i \Big/ \sum_{i}^{n} \mathbf{1}\{X_i > x_n\}, \tag{14.3}$$

where $Y_i = \ln(X_i/x_n)\mathbf{1}\{X_i > x_n\}$, numbers $\{x_n\}$ are to be chosen by a statistician. Note that Hill's estimator (14.2) is a particular case of the ratio estimator (14.3): $a_n^H = a_n$ when $x_n = X_{(k_n+1)}$.

Let $X, X_1, X_2 \ldots$ be a sequence of independent observations over the distribution of a random variable X. Denote

$$p_n = \mathbf{P}(X > x_n), \quad v_1 \equiv v_1(x_n) = \mathbf{E}\{Y|X > x_n\}/a - 1,$$

and assume that

$$x_n \to \infty, \ np_n \to \infty \qquad (n \to \infty). \tag{14.4}$$

Sufficient conditions for the asymptotic normality of the ratio estimator have been found by Goldie and Smith (1987). Novak and Utev (1990) and Novak (1992) have proved the following result.

Theorem A. *The estimator a_n is consistent:* $a_n \xrightarrow{p} a$. *The convergence*

$$\sqrt{np_n}(a_n - a) \Rightarrow \mathcal{N}(0; a^2) \tag{14.5}$$

holds if and only if $np_n v_1^2 \to 0$ *as* $n \to \infty$.

Of definite interest is the accuracy of normal approximation for a_n and the asymptotics of the mean squared error $\mathbf{E}(a_n/a - 1)^2$. The following assertion is established in Novak (1996).

Theorem B. *For all large enough n,*

$$\sup_y \left| \mathbf{P}\left(\sqrt{np_n}\left(\frac{a}{a_n} - 1\right) < y\right) - \mathit{Phi}\left(y + \frac{v_1\sqrt{np_n}}{1 + v_1}\right) \right| \le \kappa_n,$$

$$\mathbf{E}(a_n/a - 1) = v_1 + O\left((np_n^{-2})\right),$$

$$np_n \mathbf{E}(a_n/a - 1)^2 = 1 + 2(v_2 - v_1) - v_1^2 + v_1^2 np_n + \frac{1 + o(1)}{np_n}(14.6)$$

where $\kappa_n = \frac{80}{\sqrt{np_n}} + \frac{36}{np_n} + |v_2 - 3v_1|\sqrt{\frac{2}{\pi}}$ *and* $v_2 \equiv v_2(n) = a^{-2}\mathbf{E}\{Y^2|X > x_n\}/2 - 1$.

A comparison between Hill's and few similar estimators is given in de Haan and Peng (1998). A comparison between the Hill and the ratio estimators in Novak and Utev (1990) indicates that the latter seems to have an advantage. For the ratio estimator we know a necessary and sufficient condition of asymptotic normality (see Theorem A). The asymptotics of the mean squared error (in the general situation (14.1)) and a Berry–Esseen–type inequality seem to be known only for the ratio estimator.

The important question is how to choose the threshold x_n. The theoretically optimal threshold x_n^{opt} is the value x_n that minimises $v_1^2 + 1/np_n$ — the main term of $\mathbf{E}(a_n - a)^2$. Explicit expression for x_n^{opt} can be drawn under additional restrictions on the distribution (14.1).

One of important subclasses is the family

$$\mathcal{P}_{a,b,c,d} = \left\{ \mathbf{P} : \mathbf{P}(X > x) = cx^{-1/a}\left(1 + dx^{-b}(1 + o(1))\right) \right\}.$$

If $\mathbf{P} \in \mathcal{P}_{a,b,c,d}$ then, using (14.7), we get $v_1(x) \sim \frac{-bdx^{-b}}{a^{-1}+b}$. Hence $x_n^{\mathrm{opt}} = \left(2ab\left(\frac{bd}{a^{-1}+b}\right)^2 cn\right)^{\frac{a}{1+2ab}}$

Adaptive versions of x_n^{opt} may be constructed by replacing a, b, c, d with their consistent estimators $\hat{a}, \hat{b}, \hat{c}, \hat{d}$ such that $|\hat{a} - a| + |\hat{b} - b| = o_p(1/\ln n)$.

Another approach is to plot $a_n(\cdot)$ and then take the estimate from an interval where the function $a_n(\cdot)$ demonstrates stability. The background for this approach is provided by our consistency result. Indeed, if $\{x_n\}$ obeys (14.4) then so does $\{tx_n\}$ for every $t > 0$. Hence there must be an interval of threshold levels $[x_-; x_+]$ such that $a_n \approx a$ for all $x \in [x_-; x_+]$.

We simulated 1000 i.i.d. observations from the Cauchy $\mathbf{K}(0; 1)$ distribution:

The first picture says that the ratio estimator $a_n(x)$ behaves rather stable in the interval $x \in [0.5; 17]$. The second picture is even more convincing.

Observe that $\mathbf{K}(0; 1) \in \mathcal{P}_{1,1/2,1/\pi,-1/3}$. Therefore, $x_n^{\mathrm{opt}} = (16n/81\pi)^{1/5} \approx 2.29$.

Note that $|v_1| + |v_2| \to 0$ as $x_n \to \infty$. Moreover,

$$\mathbf{E}\{Y^k | X > x_n\} = a^k k! (1 + v_k), \tag{14.7}$$

where

$$v_k \equiv v_k(n) = \int_0^\infty h_n(u)e^{-u}du^k/k!, \quad h_n(u) = L^{-1}(x_n)L(x_n e^{au}) - 1.$$

Using properties of slowly varying functions, one can check that $v_k(n) \to 0$ as $x_n \to \infty$.

If $\{(\xi_i, \eta_i)\}$ are the i.i.d. pairs of random variables, $\eta_1 \geq 0$, $\mathbf{E}|\xi_1| + \mathbf{E}\eta_1 < \infty$ and $0/0 := 0$ then

$$\sum_i^n \xi_i / \sum_i^n \eta_i \xrightarrow{p} \mathbf{E}\xi_1/\mathbf{E}\eta_1.$$

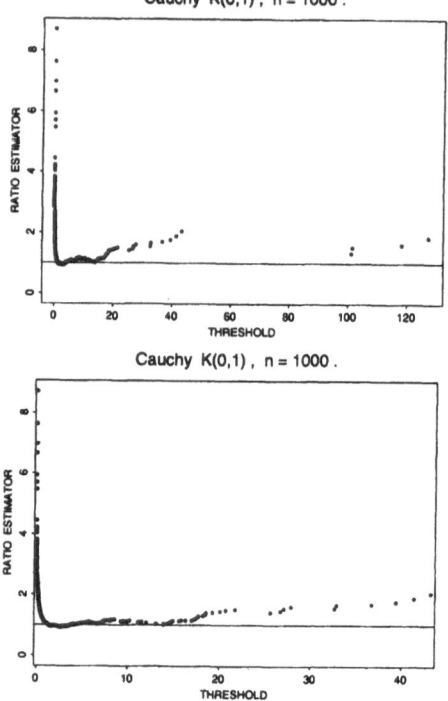

This simple fact and (14.7) form a background for a number of tail index estimators. Denote $1_i = 1\{X_i > x_n\}$, and let

$$a_{n,m} \equiv a_{n,m}(x_n) = \frac{\sum^n Y_i^m}{m! \sum^n 1_i} . \tag{14.8}$$

Proposition 14.1.1 *There holds* $a_{n,m} \xrightarrow{p} a^m$.

This implies, in particular, that

$$a_n^* = \frac{\sum^n Y_i^2}{2 \sum^n Y_i} \tag{14.9}$$

is a consistent estimator of a: $a_n^* \xrightarrow{p} a$. One can easily check that

$$\sqrt{np_n}(a_n^* - a) \Rightarrow \mathcal{N}(\mu; 2a^2) \tag{14.10}$$

if $a\sqrt{np_n}(v_2 - v_1) \to \mu$.

The estimator $a_n^* \equiv a_n^*(x_n)$ is a generalisation of de Vries' estimator

$$a_n^V = \sum_{i=1}^{k_n} \ln^2 \left(X_{(i)}/X_{(k_n+1)} \right) \left/ 2 \sum_{i=1}^{k_n} \ln \left(X_{(i)}/X_{(k_n+1)} \right) \right. .$$

Indeed, $a_n^v = a_n^*$ if we choose $x_n = X_{(k_n+1)}$.

Denote $\lambda_n = \sqrt{1 + 2v_2 - 2v_1 - v_1^2} \Big/ \left[(1 + v_1)\sqrt{1 - p_n}\right]$, $y_n = \frac{1}{3}\sqrt{np_n}\,\lambda_n$, and

$$\tilde{a} \equiv \tilde{a}(n) = \mathbf{E}\{Y|X > x_n\}.$$

The constant $C_* \leq 0.8$ is that from the Berry–Esseen inequality (see Novak (1998)).

Theorem 14.1.1 *If* $|y| \leq y_n$ *then*

$$\Delta_n(y) \equiv |\mathbf{P}\left(\sqrt{np_n}(a_n/\tilde{a} - 1) < y\right) - \Phi(y)|$$

$$\leq \frac{4C_*}{\sqrt{np_n}} + \frac{|2v_1 - v_2|}{\sqrt{2\pi e}} + O\left(\frac{1}{np_n} + \sum_1^3 v_i^2\right) \quad (14.11)$$

as $n \to \infty$. *If* $|y| > y_n$ *then* $\Delta_n(y) \leq 16/np_n$.

In fact, we estimate $\tilde{a} = a(1 + v_1(n))$ instead of a. The justification is that

$$\mathbf{P}(X > x) \sim L(x)x^{-1/\tilde{a}} \quad (14.12)$$

uniformly in $x \in [1; e^{t_n}]$ for any sequence $\{t_n\}$ such that $t_n = o(1/v_1)$. According to (14.6),

$$\mathbf{E}\left(a_n/\tilde{a} - 1\right)^2 \sim v_1^2 + 1/np_n \quad (n \to \infty)$$

(the so–called bias–versus–variance trade–off). Thus, it would be appropriate to choose numbers $\{x_n\}$ so that $v_1 = O(1/\sqrt{np_n})$. If so, (14.12) holds for a large enough interval of values of x.

Denote

$$\nabla_n(y) = \mathbf{P}\left(\sqrt{\sum 1_i}\left(\frac{a_n}{\tilde{a}} - 1\right) < y\right) - \Phi(y).$$

Theorem 14.1.2 *If* $|y| \leq y_n$ *then, as* $n \to \infty$,

$$\nabla_n(y) \leq \frac{4C_*}{\sqrt{np_n}} + \frac{|2v_1 - v_2|}{\sqrt{2\pi e}} + O\left(\frac{1}{np_n} + \sum_1^3 v_i^2\right), \quad (14.13)$$

$$\nabla_n(y) \geq -\frac{4C_*}{\sqrt{np_n}} - \frac{|2v_1 - v_2|}{\sqrt{2\pi e}} + O\left(\frac{\ln(np_n)}{np_n} + \sum_1^3 v_i^2\right)(14.14)$$

Thus, one can suggest for $\tilde{a}$ the confidence interval

$$I = \left[a_n \left(1 + \frac{y}{\sqrt{\sum 1_i}} \right)^{-1} ; \, a_n \left(1 - \frac{y}{\sqrt{\sum 1_i}} \right)^{-1} \right] \qquad (14.15)$$

of the level $1 - q$, where y is the q^*-quantile of Φ,

$$q^* = \left(\frac{q}{2} - \frac{4C_*}{\sqrt{\sum 1_i}} - \frac{|2v_1^* - v_2^*|}{\sqrt{2\pi e}} \right)_+, \, v_1^* \text{ and } v_2^* \text{ are estimates of } v_1 \text{ and } v_2.$$

The asymptotic approach to constructing confidence intervals can yield a wrong answer if the sample size is not large and the rate of convergence in a CLT for a tail index estimator is slow. Apart from asymptotic confidence intervals, (14.15) is constructed using a Berry–Esseen–type inequality. It hence takes into account the rate of convergence in the corresponding CLT.

Besides heavy tails, financial data often exhibits dependence. Thus, it is important to develop procedures of the tail index estimation from sequences of weakly dependent random variables. Assume that $X, X_1, X_2 \ldots$ is a (strictly) stationary sequence of of random variables. The definitions of the weak dependence coefficients α, φ and ρ can be found, eg., in Bradley (1986), Utev (1989). Remind that $\rho(\cdot) \leq 2\sqrt{\varphi(\cdot)}$.

Hsing (1991) and Resnick and Starica (1998) suggested sufficient conditions for consistency of Hill's estimator in the case of m–dependent sequences and some classes of stationary processes. Complicated sufficient conditions for asymptotic normality of Hill's estimator for dependent data are given by ? and Drees (1999).

We show that the ratio estimator is consistent under rather simple assumptions expressed in terms of the φ–mixing coefficient.

Theorem 14.1.3 *If $\sum_{i \geq 0} \rho(2^{i/3}) < \infty$ then the ratio estimator is consistent:* $a_n \xrightarrow{p} a$.

Condition $\sum_{i \geq 0} \rho(2^{i/3}) < \infty$ is rather weak: in many models $\varphi(\cdot)$ (and, therefore, $\rho(\cdot)$) decays exponentially fast (cf. Davis, Mikosch and Basrak (1999)).

Proof of Theorem 4. We use Chebyshev's inequality, (14.7) and an estimate of a variance of a sum of dependent random variables (see Peligrad (1982) or Utev (1989)). For any $\varepsilon > 0$,

$$\begin{aligned} \mathbf{P}(a_n - \tilde{a} > \varepsilon) &= \mathbf{P}\left(\sum^n (Y_i - \tilde{a}) 1_i > \varepsilon \sum^n 1_i \right) \\ &= \mathbf{P}\left(\sum^n Z_i > \varepsilon n p_n \right) \leq (\varepsilon n p_n)^{-2} \mathrm{Var}\left(\sum^n Z_i \right), \end{aligned}$$

where $Z_i = (Y_i - \tilde{a})\mathbf{1}_i - \varepsilon(\mathbf{1}_i - p_n)$. By Theorem 1.1 in Utev (1989), there exists a constant c_ρ (depending only on $\rho(\cdot)$) such that $\mathrm{Var}\left(\sum^n Z_i\right) \leq c_\rho n \mathrm{Var}\, Z_1 \leq cn p_n$ (we used also (14.7)). Hence $\mathbf{P}(a_n - \tilde{a} > \varepsilon) \to 0$. Similarly we check that $\mathbf{P}(a_n - \tilde{a} < -\varepsilon) \to 0$. Remind that $\tilde{a} \to a$ as $x_n \to \infty$. The result follows.

Bibliography

Bradley, R. (1986). Basic properties of strong mixing condition, *in* E.Eberlein and M.S.Taqqu (eds), *Dependence in Probability and Statistics*, Boston: Birkhäuser, pp. 165–192.

Csörgő, S. and Viharos, L. (1998). *Asymptotic Methods in Probability and Statistics*, Elsevier, chapter Estimating the tail index, pp. 833–881.

Davis, R., Mikosch, T. and Basrak, B. (1999). Sample acf of multivariate stochastic recurrence equations with applications to garch, *Technical report*, University of Groningen, Department of Mathematics.

de Haan, L. and Peng, L. (1998). Comparison of tail index estimators, *Statistica Neerlandica* **52**(1): 60–70.

Drees, H. (1999). Weighted approximations of tail processes under mixing conditions, *Technical report*, University of Cologne.

Embrechts, P., Klüppelberg, C. and Mikosch, T. (1997). *Modelling Extremal Events for Insurance and Finance*, Springer Verlag.

Hsing, T. (1991). On tail index estimation for dependent data, **19**(3): 1547–1569.

Novak, S. (1996). On the distribution of the ratio of sums of random variables, *Theory Probab. Appl.* **41**(3): 479–503.

Novak, S. (1998). Berry–esseen inequalities for a ratio of sums of random variables, *Technical Report 98*, University of Sussex.

Novak, S. and Utev, S. (1990). On the asymptotic distribution of the ratio of sums of random variables, *Siberian Math. J.* **31**: 781–788.

Resnick, S. (1997). Heavy tail modeling and teletraffic data, *Ann. Statist.* **25**(5): 1805–1869.

Resnick, S. and Starica, C. (1998). Tail index estimation for dependent data, *Ann. Appl. Probab.* **8**(4): 1156–1183.

Utev, S. (1989). Sums of φ–mixing random variables, *Technical report*, Trudy Inst. Mat. (Novosibirsk).

15 Extremes of alpha-ARCH Models

Christian Robert

15.1 Introduction

In the recent literature there has been a growing interest in nonlinear time series models. Many of these models were introduced to describe the behavior of financial returns. Large changes tend to be followed by large changes and small changes by small changes (see Mandelbrot (1963)). These observations lead to models of the form $X_t = \sigma_t \varepsilon_t$, where the conditional variance depends on past information.

Two approaches have been proposed to specify time-dependent variances. The first one assumes that the conditional variances are generated by a nonlinearly transformed underlying stochastic process, for example an autoregressive process. These models are called stochastic volatility (SV) models. In the second approach, the variance of the series is a deterministic function of lagged observations like in the ARCH-GARCH models (Engle (1982), Bollerslev (1986)). For a review with financial applications, see Bollerslev, Chou and Bollerslev, Chou and Kroner (1992) or Gourieroux (1997).

The extremal behavior of a process $(X_t)_{t \in \mathbb{N}}$ is for instance summarized by the asymptotic behavior of the maxima:

$$M_n = \max_{1 \le k \le n} X_k, \ n \ge 1.$$

The limit behavior of M_n is a Markov chain, is a well-studied problem in extreme value theory, when $(X_t)_{t \in \mathbb{N}}$ (see e.g. Leadbetter and Rootzen (1988) or Perfect (1994)). See also Leadbetter, Lindgren and Rootzen (1983) and the references therein, for a general overview of extremes of Markov processes. Under general mixing conditions, it can be shown that for large n and x:

$$P(M_n \le x) \approx F^{n\theta}(x),$$

where F is the stationary distribution of $(X_t)_{t \in \mathbb{N}}$ and $\theta \in (0, 1]$ is a constant called *extremal index*. This index is a quantity which characterizes the relationship between the dependence structure of the data and their extremal behavior. It can be shown that exceedances of a high threshold value tend to occur in clusters whose average size is related to the inverse of θ.

Extremes of special gaussian stochastic volatility models have been studied by Breidt and Davis (1998). They showed that the law of the normalized maximum of the logarithm of the square of the process converge to the Gumbel law. Extremal behavior for the ARCH(1)-model is described in Haan, Resnick, Rootzen and Vries (1989), for the GARCH(1,1)-model in Mikosch and Starica (1998), and for the general GARCH(p,q)-model in Davis, Mikosch and Basrak (1999). They get Pareto-like tail and extremal index strictly less than one.

The purpose of this article is to study autoregressive conditional heteroskedastic models with a similar structure as the classical ARCH-model, and to show that they feature Weibull-like tails and extremal index equal to one. These models are the α-ARCH-models introduced by Diebolt and Guegan (1991). Recent empirical works (Cont, Potters and Bouchaud (1997),GGourieroux, Jasiak and Le Fol (1999), Robert (2000) have shown that some financial series could exhibit such behaviors. These series are mainly high frequency data.

This paper is organized as follows. In section 2, we present the model and we recall its main properties. In section 3, we determine the tails of the stationary distribution. In section 4, we show that the extremal index exists and is equal to one, and we give the law of the asymptotic normalized maximum. Section 5 is an application on an high frequency series of the price variations for the Alcatel stock traded on the Paris Bourse. The proofs are gathered in Section 6. Section 7 concludes the paper.

15.2   The model and its properties

The model is defined by the autoregressive scheme:

$$X_t = \sqrt{a + b\left(X_{t-1}^2\right)^\alpha}\,\varepsilon_t, \; t \in \mathbb{Z}, \tag{15.1}$$

where a, b are positive constants and $0 < \alpha < 1$. $\{\varepsilon_t\}_{t \in \mathbb{Z}}$ is a sequence of i.i.d. random variables with standard gaussian distribution[1]. This is

---

[1]We assume a common gaussian distribution for the ε_t. We can introduce

the Diebolt and Guegan (1991) model with autoregressive conditional heteroskedastic errors of order one and as autoregressive part.

When $\alpha = 0$, the model reduces to a sequence of i.i.d. gaussian random variables, and when $\alpha = 1$, it reduces to the classical ARCH(1) model with gaussian innovations. Extremal behaviors of these two processes are well known [see, for example, Resnick (1987) p.71 for the first case, and Haan et al. (1989) for the second one].

The process is a homogeneous Markov process with state space $\mathbb{R}$. The transition density is given by:

$$P(X_1 \in [u, u + du]|X_0 = x) = \frac{1}{\sqrt{2\pi}} \frac{1}{\sqrt{a + bx^{2\alpha}}} \exp\left(-\frac{u^2}{2(a + bx^{2\alpha})}\right) du,$$

with $x \in \mathbb{R}$.

The α-ARCH process has the following properties (Diebolt and Guegan (1991) Theorems 1 and 2):

Theorem 15.2.1 *Under the condition* $0 < \alpha < 1$:
(i) There exists a unique stationary process $(X_t)_{t \in \mathbb{Z}}$ *which is measurable with respect to the* σ-algebra $\sigma(\varepsilon_t, \varepsilon_{t-1}, ...)$ *and satisfies the recursive equation* $(15.1)^2$.
(ii) $(X_t)_{t \in \mathbb{Z}}$ *is geometrically strong mixing.*
(iii) If $\alpha > 1/2$, *then* X_t *has no exponential moment, but admits moments of any orders.*

Let us denote by F its stationary distribution. A random variable X with distribution F satisfies the fixpoint equation:

$$X \overset{d}{=} \sqrt{a + b(X^2)^\alpha}\varepsilon, \tag{15.2}$$

where ε is a standard gaussian random variable, independent of X.

Let:

$$h(x) = \frac{1}{\sqrt{2\pi}}\exp\left(-\frac{x^2}{2}\right), \quad \bar{H}(x) = \frac{1}{\sqrt{2\pi}}\int_x^\infty \exp\left(-\frac{u^2}{2}\right) du,$$

be the density and the survivor function of the standard normal distribution, f the density of the distribution F.

---

weaker assumptions by considering distributions with densities such $h(x) = C|x|^\beta \exp(-k|x|^\gamma)$, $\beta \in \mathbb{R}$ and $\gamma > 0$.
[2]Moreover, $\sigma(\varepsilon_t, \varepsilon_{t-1}, ...) = \sigma(X_t, X_{t-1}, ...)$, and then ε_t is independent of $\sigma(X_t, X_{t-1}, ...)$.

We have:

$$f(x) = \int_{-\infty}^{\infty} \frac{1}{\sqrt{a + bt^{2\alpha}}} h\left(\frac{x}{\sqrt{a + bt^{2\alpha}}}\right) f(t)dt, \ \forall x \in \mathbb{R}.$$

Since the function h is symmetric, the pdf f is also symmetric, and we have:

$$f(x) = 2\int_{0}^{\infty} \frac{1}{\sqrt{a + bt^{2\alpha}}} h\left(\frac{x}{\sqrt{a + bt^{2\alpha}}}\right) f(t)dt, \ \forall x \in \mathbb{R}. \qquad (15.3)$$

15.3   The tails of the stationary distribution

In this section, we study the tails of the stationary distribution. Diebolt and Guegan (1991) derived conditions for the existence of polynomial or exponential moments. Here we completely study the pattern of the tails for $\alpha > 1/2$.

First, we associate an auxiliary model to model (15.1). Following Breidt and Davis (1998) , we get for this model the pattern of the tail distribution, and then deduce the pattern of the tail for the initial model.

Let first assume that $a = 0$, then we obtain:

$$Y_t = \alpha Y_{t-1} + U_t, \qquad (15.4)$$

where $Y_t = \ln X_t^2$ and $U_t = \ln(b\varepsilon_t^2)$. This is an autoregressive model of order one (AR(1)). Since $0 < \alpha < 1$ and U_t admits a positive density on $\mathbb{R}$, the existence and the uniqueness of a stationary distribution F_Y is guaranteed. A random variable Y with distribution F_Y satisfies the fixpoint equation:

$$Y \overset{d}{=} \alpha Y + U. \qquad (15.5)$$

U is a random variable with distribution has as $\ln(b\varepsilon^2)$, where ε is standard normal, independent of Y.

The transformation of the process guarantees the existence of exponential moments (see (iii) of Theorem 15.2.1). Let us now introduce the logarithm of the moment-generating function and its first two derivatives:

$$\begin{aligned}
q_0(\lambda) &= \ln C_0(\lambda) = \ln E(\exp\{\lambda Y\}), \\
m_0(\lambda) &= \frac{d}{d\lambda} q_0(\lambda), \\
S_0^2(\lambda) &= \frac{d^2}{d\lambda^2} q_0(\lambda).
\end{aligned}$$

The latter will be useful to find the tail pattern of Y.

¿From (15.5), we have:

$$
\begin{aligned}
q_0(\lambda) &= q_0(\alpha\lambda) + \ln E(\exp\{\lambda U\}), & (15.6)\\
&= q_0(\alpha\lambda) + \lambda \ln 2b + \ln\Gamma(1/2 + \lambda) - \ln\Gamma(1/2),\\
&= q_0(\alpha\lambda) + \lambda\ln\lambda + \lambda(\ln 2b - 1) + \ln 2/2 + O(1/\lambda). & (15.7)
\end{aligned}
$$

Then we deduce an asymptotic expansion of the function q_0 (see Appendix 1):

$$
q_0(\lambda) = \frac{\lambda\ln\lambda}{(1-\alpha)} + \frac{\left[\ln\left(2b\alpha^{\frac{\alpha}{1-\alpha}}\right) - 1\right]}{(1-\alpha)}\lambda - \frac{\ln 2}{2\ln\alpha}\ln\lambda + m + O(1/\lambda),
$$
$$(15.8)$$

where m is a constant. We can also derive asymptotic expansions of its derivatives:

$$
\begin{aligned}
m_0(\lambda) &= \frac{\ln\lambda}{(1-\alpha)} + \frac{\ln\left(2b\alpha^{\frac{\alpha}{1-\alpha}}\right)}{(1-\alpha)} - \frac{\ln 2}{2\ln\alpha}\frac{1}{\lambda} + O(1/\lambda^2), & (15.9)\\
S_0^2(\lambda) &= \frac{1}{\lambda(1-\alpha)} + \frac{\ln 2}{2\ln\alpha}\frac{1}{\lambda^2} + O(1/\lambda^3). & (15.10)
\end{aligned}
$$

Let:

$$
m_0^{-1}(x) = \frac{\alpha^{-\frac{\alpha}{1-\alpha}}}{2b}e^{(1-\alpha)x} + \frac{(1-\alpha)\ln 2}{2\ln\alpha} + O\left(e^{-(1-\alpha)x}\right); \qquad (15.11)
$$

we get:

$$
m_0\left\{m_0^{-1}(x)\right\} = x + O\left(e^{-(1-\alpha)x}\right),
$$

which justifies the notation inverse $^{-1}$.

Theorem 15.3.1 *The stationary distribution of the process (Y_t) defined in (15.4) satisfies:*

$$
\begin{aligned}
\bar{F}_Y(y) &= P[Y > y] \underset{y\to\infty}{\sim} \frac{\exp\left\{-ym_0^{-1}(y)\right\}C_0\left\{m_0^{-1}(y)\right\}}{m_0^{-1}(y)S_0(m_0^{-1}(y))(2\pi)^{1/2}} & (15.12)\\
&= 2D\exp\left\{-\frac{\alpha^{-\frac{\alpha}{1-\alpha}}}{2b(1-\alpha)}e^{(1-\alpha)y} - y\frac{(1-\alpha)}{2}\left[\frac{\ln 2}{\ln\alpha}+1\right]\right.\\
&\qquad\left. + O\left(ye^{-(1-\alpha)y}\right)\right\},
\end{aligned}
$$

where D is a positive constant[3].

A similar result can be deduced when $a > 0$.

Theorem 15.3.2 *The stationary distribution of the process (X_t) defined in (15.1) satisfies:*

$$
\begin{aligned}
\bar{F}_X(x) &= P[X > x] \\
&= D\exp\left\{-\frac{\alpha^{-\frac{\alpha}{1-\alpha}}}{2b(1-\alpha)}x^{2(1-\alpha)} - (1-\alpha)\left[\frac{\ln 2}{\ln \alpha} + 1\right]\ln x \right. \\
&\qquad\left. + O\left(x^{-2(1-\alpha)}\ln x\right)\right\}.
\end{aligned}
$$

if and only if $1/2 < \alpha < 1$.

We can now study the type of tail of the variable X. Let us first introduce the definition of a Weibull-like tail.

Definition 15.3.1 *A random variable has a Weibull-like tail if its survivor function is:*

$$
\bar{F}(x) = \exp\left\{-cx^{\tau}l(x)\right\},
$$

where c, τ are strictly positive constants and:

$$
\lim_{x\to\infty} l(x) = 1.
$$

X has a Weibull-like tail, since its survivor function satisfies[4]:

$$
\bar{F}_X(x) \underset{x\to\infty}{\sim} \frac{D}{x^f}\exp\left\{-ex^c\right\}.
$$

with:

$$
c = 2(1-\alpha), \qquad e = \frac{\alpha^{-\frac{\alpha}{1-\alpha}}}{2b(1-\alpha)}, \qquad f = (1-\alpha)\left[\frac{\ln 2}{\ln \alpha} + 1\right].
$$

---

[3]Let m and n be two functions. $m(y) \underset{y\to\infty}{\sim} n(y)$ if and only if $\lim_{y\to\infty} m(y)/n(y) = 1$.

[4] By a scale change:

$$
\lambda X_t = \sqrt{\lambda^2 a + b\lambda^{2(1-\alpha)}(\lambda X_{t-1})^{2\alpha}}\,\varepsilon_t, \qquad \lambda > 0,
$$

and only the coefficients D and e are changed. The tail pattern is not changed.

Moreover, the density of the stationary distribution f satisfies:

$$f(x) \underset{x \to \infty}{\sim} \frac{D}{x^d} \exp(-ex^c), \text{ avec } d = \alpha + \frac{(1-\alpha)}{\ln(\alpha)} \ln 2.$$

We completely specified the tails, eventhough the constant D is not explicit. In the ARCH-model, this parameter is known, but is not easily expressed in terms of structural parameters.

15.4 Extreme value results

In this section we propose normalizing factors which ensure that the normalized maxima of an i.i.d. sequence with the same common distribution as the stationary one converge to a non-degenerate distribution. Then, we check that the maxima of the stationary sequence exhibit the same behavior. We suppose that the distribution of X_0 is the stationary one.

15.4.1 Normalizing factors

Let us write the survivor function $\bar{F}(x)$ as:

$$\begin{aligned} \bar{F}(x) &= D \exp\left(-ex^c - f \ln(x)\right) h_F(x), \\ \bar{F}(x) &= AD \exp\left(-\int_1^x \frac{1}{a(t)} dt\right) h_F(x), \end{aligned}$$

with:

$$a(t) = \frac{1}{ect^{c-1} + \frac{f}{t}}, \qquad A = \exp(-ec - f), \qquad \lim_{x \to \infty} h_F(x) = 1.$$

According to propositions 1.1 and 1.4 of Resnick (1987), F belongs to the maximum domain of attraction of the Gumbel law. Moreover, the normalizing factors which ensure that:

$$F^n(c_n x + d_n) \underset{n \to \infty}{\to} \Lambda(x) = \exp\left\{-e^{-x}\right\},$$

can be chosen such as:

$$-\ln \bar{F}(d_n) = \ln n, \qquad c_n = a(d_n).$$

Theorem 15.4.1 *If* $\left(\tilde{X}_t\right)_{t \in \mathbb{N}}$ *is an i.i.d. sequence of random variables with common distribution F, then:*

$$P\left\{\left(Max\left(\tilde{X}_1, ..., \tilde{X}_n\right) - d_n\right) / c_n \leq x\right\} \underset{n \to \infty}{\to} \Lambda(x),$$

with:

$$d_n = \left(\frac{\ln n}{e}\right)^{1/c} + \frac{1}{c}\left(\frac{\ln n}{e}\right)^{1/c-1}\left(-\frac{f}{ec}\ln\left(\frac{\ln n}{e}\right) + \frac{\ln D}{e}\right),$$

$$c_n = \frac{1}{ec}\left(\frac{\ln n}{e}\right)^{1/c-1}.$$

15.4.2   Computation of the extremal index

Finally, we would like to know if the extremal index of $(X_t)_{t \in \mathbb{N}}$ exists and to compute it. After preliminary definitions, we show that the extremal index exists and is equal to one. Let (u_n) be a deterministic real sequence. For each l with $1 \leq l \leq n - 1$, define:

$$\alpha_{n,l} = \sup|P\left(X_i \leq u_n, i \in A \cup B\right) - P\left(X_i \leq u_n, i \in A\right)P\left(X_i \leq u_n, i \in B\right)|$$

where the supremum is taken over all A and B such that:

$$A \subset \{1, ..., k\}, B \subset \{k + l, ..., n\},$$

for some k with $1 \leq k \leq n - l$.

Definition 15.4.1 *Let $(u_n)_{n \in \mathbb{N}}$ be a real sequence. The condition $D(u_n)$ is said to hold if $\alpha_{n,l_n} \to 0$ as $n \to \infty$ for some sequence $l_n = o(n)$.*

This condition is implied by the strong mixing condition (see Leadbetter and Rootzen (1988) p. 442).

Definition 15.4.2 *Assume that the condition $D(u_n)$ for the stationary sequence $(Y_t)_{t \in \mathbb{N}}$. The condition $D'(u_n)$ (called "anti-clustering condition") holds if there exist sequences of integers $(s_n)_{n \in \mathbb{N}}$ and $(l_n)_{n \in \mathbb{N}}$ such that $s_n \to \infty$, $s_n l_n / n \to 0$, $s_n \alpha_{n,l_n} \to 0$, and:*

$$\lim_{n \to \infty} n \sum_{j=1}^{p_n} P\{Y_0 > u_n, Y_j > u_n\} = 0,$$

where $p_n = [n/s_n]$.

Theorem 15.4.2 *(Leadbetter et al. (1983): Assume that the conditions $D(u_n)$ and $D'(u_n)$ for a stationary sequence $(Y_t)_{t\in\mathbb{N}}$ and for a sequence $(u_n)_{n\in\mathbb{N}}$ such that $\lim_{n\to\infty} n(1 - F_Y(u_n)) = \tau$, then:*

$$\lim_{n\to\infty} P\{Max\,(Y_1, ..., Y_n) \le u_n\} = e^{-\tau}.$$

In a first step, we focus on the squared process:

$$Z_t = X_t^2 = (a + bZ_{t-1}^{\alpha})\epsilon_t^2.$$

We check that the conditions $D(u_n^Z)$ and $D'(u_n^Z)$ hold for the stationary sequence $(Z_t)_{t\in\mathbb{N}}$.

Theorem 15.4.3 *The extremal index of $(Z_t)_{t\in\mathbb{N}}$ exists and is equal to one.*

It follows that the extremes of the square of the process are nearly "independent". In a second step, we show that the extremal index of $(X_t)_{t\in\mathbb{N}}$ is also equal to one. Note that $\{X_t\} = \left\{B_t\sqrt{X_t^2}\right\}$ where the random variables $\{B_t\}$ are i.i.d., independent of $\{|X_t|\}$ and such that $P(B_1 = 1) = P(B_1 = -1) = 1/2$.

Let $(u_n(\tau))$ be such that $\lim_{n\to\infty} n(1 - F(u_n(\tau))) = \tau$, we define $N_n = \sum_{i=1}^{n} 1_{\{X_i^2 > u_n^2(\tau)\}}$, i.e. the number of times where $\{X_t^2\}$ exceeds the threshold $u_n^2(\tau)$. Let $1 \le \vartheta_1 < \vartheta_2 < ...$ be the random times where $\{X_t^2\}$ exceeds $u_n^2(\tau)$. The following conditions are equivalent:

$$\lim_{n\to\infty} n(1 - F(u_n(\tau))) = \tau,$$
$$\lim_{n\to\infty} n(1 - F_Z(u_n^2(\tau))) = 2\tau,$$
$$\lim_{n\to\infty} P\{M_n^Z \le u_n^2(\tau)\} = e^{-2\tau}.$$

Since the extremal index of $(Z_t)_{t\in\mathbb{N}}$ is equal to one, the random variable N_n converges weakly to a Poisson random variable N with parameter 2τ (see Leadbetter and Rootzen (1988)).

Then we have:

$$
\begin{aligned}
P\left\{\max(X_1, ..., X_n) \le u_n(\tau)\right\} &= \sum_{k=0}^{\infty} P\left\{N_n = k, B_{\vartheta_1} = ... = B_{\vartheta_k} = -1\right\} \\
&= \sum_{k=0}^{\infty} P\left\{N_n = k\right\} 2^{-k} \\
&\to \sum_{k=0}^{\infty} P\left\{N = k\right\} 2^{-k} \\
&= e^{-2\tau} \sum_{k=0}^{\infty} 2^{-k} \frac{(2\tau)^k}{k!} = e^{-\tau}.
\end{aligned}
$$

If $u_n(x) = c_n x + d_n$, then $\lim_{n \to \infty} n(1 - F(u_n(x))) = -\ln \Lambda(x) = e^{-x}$ and finally:

$$
\lim_{n \to \infty} P\left\{\max(X_1, ..., X_n) \le u_n(x)\right\} = \Lambda(x).
$$

15.5   Empirical study

The development of financial markets resulting in the set up of automatic quotation systems, created access to large data-sets of tick-by-tick observations. In this section, we estimate the tails and the extremal index of an high frequency series of the price variations.

We consider a set of intra-trade price variations for the Alcatel Stock over the period ranging from May 1997 to September 1998[5]. This stock is one of the most heavily traded asset. The Paris Stock Exchange operates over five days a week during seven hours per day. The original records include also observations collected before the market opening. We delete the times between the market closures and the next day openings as well as the weekend gaps. The opening and the closure take place at 10:00 a.m. and 5:00 p.m. respectively.

The main features of tick-by-tick data are the following ones:

- the transactions occur at random times.

- returns and traded volumes exhibit strong intraday seasonalities.

- the market prices take discrete values with a minimum price variation called the tick. The distribution of these variations between consecutive trades has a discrete support.

---

[5] extracted from the records of the Paris Stock Exchange (SBF Paris Bourse)

The first and third characteristics of tick by tick data can be partly neglected due to our sampling scheme. Indeed, we sampled every 30 mn from 10.15 a.m. to 16.45 p.m.. It also allows to eliminate the opening and closure effects.

We provide in Table 1 summary statistics of the series of the price variations (Δp_t), and in Figure (15.1) empirical marginal distribution.

Mean	$3.84 \cdot 10^{-2}$
Variance	14.40
Skewness	0.10
Kurtosis	9.19
Min	-24
Max	29

Table 1 : Summary statistics of price variations, sampled at 30 mn

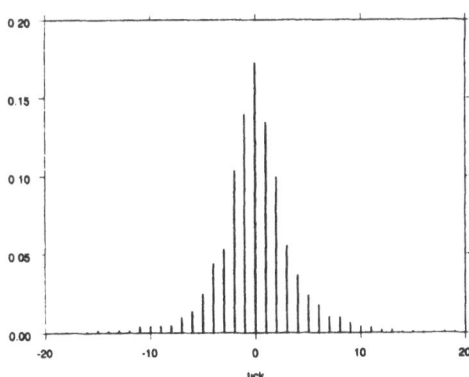

Figure 15.1: Marginal distribution

The distribution is quite symmetric. A marginal overdispersion can be exhibited, since the kurtosis is greater than 3: the tails are heavier than the normal ones.

The sample autocorrelations of the absolute values are significantly different from zero for large lags. However, it might be misleading to conclude to an empirical evidence of volatility persistence, since the hyperbolic decay of the autocorrelogram can be due to the omitted intraday seasonality. The presence of non-linearities is clear when we compare the autocorrelograms on the series and its absolute values. As a consequence, a stochastic volatility model or an α-ARCH model seem to be appropriate.

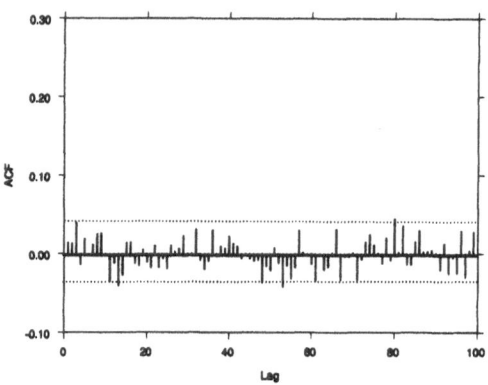

Figure 15.2: Sample ACF of price variations

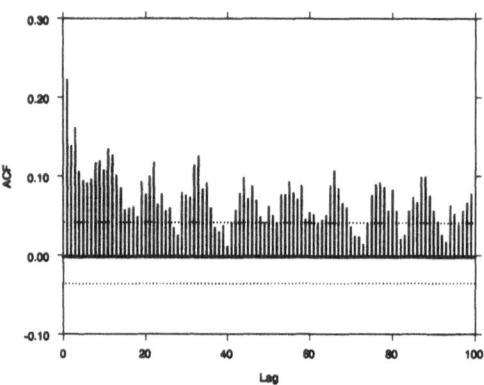

Figure 15.3: Sample ACF of absolute values of the price variations

15.5.1 Distribution of extremes

We define as extreme price variation, as any price variation that exceeds 6 ticks.

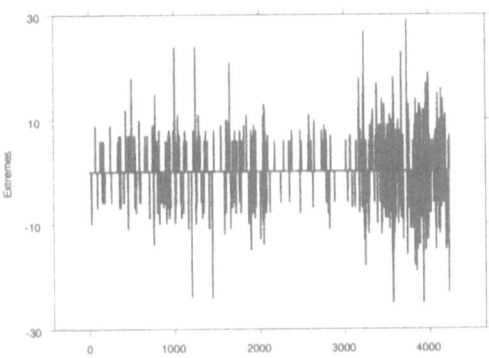

Figure 15.4: Extreme variations over the period May 1997 and September 1998

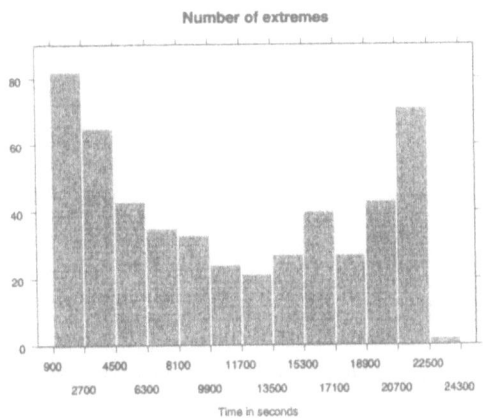

Figure 15.5: Distribution of extremes during a day

The series of extremes is given in the Figure (15.4). Large movements usually take place during the last four months of the sample, on and after June 1998. We also note the clustering of extremal events: a large price movement is often followed by large pices movements. Therefore, the extremal index could be less than one.

Extreme price movements usually occur during the first hours of the trading session and right before the closure (see Figure (15.5)). A simple explanation for this phenomenon is the market activity during these hours.

15.5.2   Tail behavior

Let us now assume that $\Delta p_t = [X_t]$, where $(X_t)_{t \in \mathbb{N}}$ is an α-ARCH process, and $[a]$ the integer part of a. We approximate the latent process (X_t) by $Y_t = \Delta p_t + U_t - 0.5$, where (U_t) are independent random variables with uniform distribution and are independent of (Δp_t). This transformation is applied to get data with continuous values. IT is easily seen that the distributions of X_t and Y_t have the same tails (see Appendix 2).

Various methods can be used to analyze the tails. Gumbel stresses the importance of looking at data before engaging in a detailed statistical analysis. For this reason we first look at the Quantile-Quantile Plot. If the data are generated from a random sample of the benchmark distribution, the plot looks roughly linear. This remains true if the data come from a linear transformation of the distribution. Some difference in the distributional patterns may be also deduced from the plot. For example if the benchmark distribution has a lighter upper-tail than the true distribution, the Q-Q plot will curve down at the right of the graph.

In Figure (15.6), we compare the upper-tail of the distribution of Y_t with the Weibull one, and in Figure (15.7), we compare with the Pareto one. Hence, we can see that the upper-tail is Weibull-like, which confirms the empirical study in Robert (2000).

To estimate the value of τ, the Weibull coefficient, a method considers the inverse of the slope of the line of the Figure (15.6). We obtain $\tau \simeq 0.80$.

A second estimator has been proposed by Beirlant, Broniatowski, Teugels

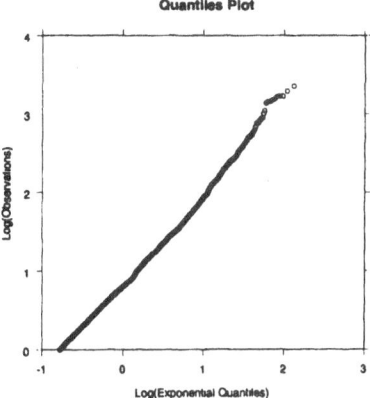

Figure 15.6: Comparison between the upper-tails of the empirical and Weibull distributions

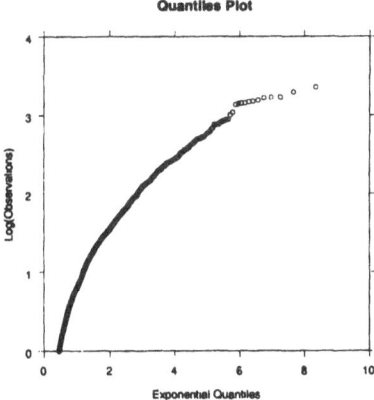

Figure 15.7: Comparison between the upper-tails of the empirical and Pareto distributions

and Vynckkier (1995):

$$B_{k,n}^X = \frac{\log(n/k)}{X_{(k)}} \frac{1}{k} \sum_{i=1}^{k} \left(X_{(i)} - X_{(k+1)} \right),$$

where $X_{(k)}$ is the kth largest order statistic of $X_1, X_2, ..., X_n$.

We know that $1/B_{k,n}^X \xrightarrow{P} \tau$ when the data are independent. It is shown in Robert [1999b] that for a α-ARCH process, the estimator is still

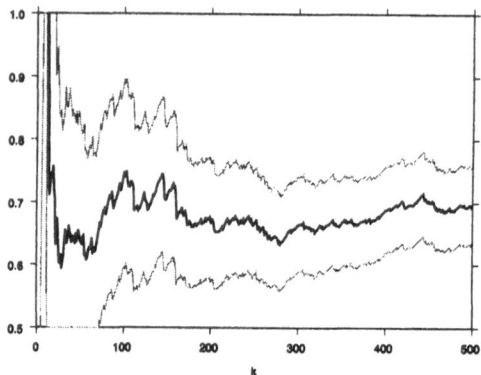

Figure 15.8: Estimator BBTV $(1/B_{k,n}^Y)$

consistent to $\tau = 2(1 - \alpha)$.

In Figure (15.8), we plot the these estimators according to k, and their asymptotic confidence bound corresponding to the independent case. We obtain $\hat{\tau} \approx 0.7$.

15.5.3   The extremal index

Various methods have been proposed in the literature to estimate the extremal index (cf. Embrechts, C. and Mikosch (1997), Section 8.1). We use below the so-called blocks method which divides the set of N observations into K blocks of length n. Each block can be considered as a cluster of exceedances. Then, we can select a threshold u. Two quantities are of interest: the number K_u of blocks in which at least one exceedance of the threshold u occurs, and the total number N_u of exceedances of u.

A natural estimator of θ is:

$$\hat{\theta} = \frac{K}{N} \frac{\ln(1 - K_u/K)}{\ln(1 - N_u/N)}.$$

Under general conditions, $\hat{\theta}$ is consistent (Smith and Weismann (1994)).

On Figure (15.9), we plot the estimates of the extremal index for different threshold u (on the x-axis) and for different lengths n of the blocks.

The estimated extremal index should be taken less than one ($\hat{\theta} \approx 0.75$), whereas the unitary value is associated with the α-ARCH model. This could be a consequence of the particular distribution of extremes during a trading day caused by the intraday seasonalities, but also of a slow rate of convergence of the estimator. To clarify the latter point, we first fix the value of α and estimate the coefficients of the model by maximum-likelihood. We get $\hat{a} = 1.51$ ($\hat{\sigma}_a = 0.13$) and $\hat{b} = 4.06$ ($\hat{\sigma}_b = 0.45$).

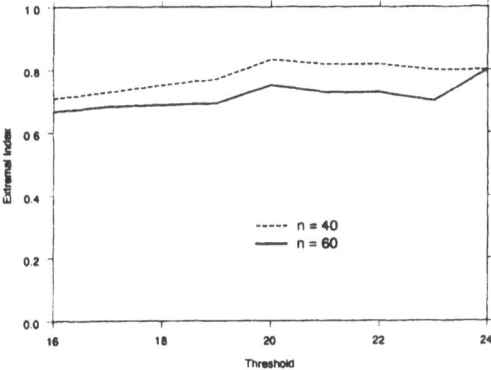

Figure 15.9: Extremal index

Then, we use this estimated α-ARCH model to simulate 100 series with the same length as the sample, and to deduce the finite sample distribution of the estimated extremal index. We choose $n = 40$ and we consider the same thresholds as before. On Figure (15.10), we plot for each threshold the mean and the 5th and 95th quantiles of the finite sample distribution.

We can see that the convergence to the true value is very slow, and the previous results can not be used to reject the α-ARCH model.

15.6 Proofs

Proof of theorem 15.3.1. The proof of this theorem is similar to the arguments given in Breidt and Davis (1998). It relies on the asymptotic normality for the normalized Escher transform.

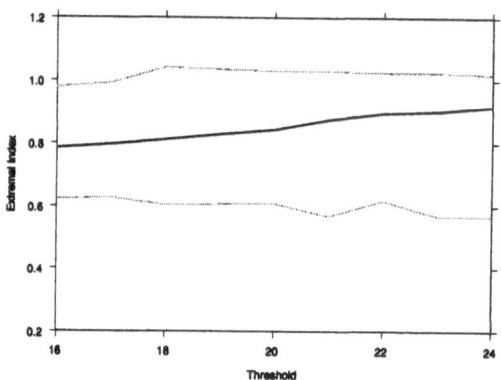

Figure 15.10: Simulated extremal index

Consider the family of probability density functions defined by:

$$g_\lambda(x) = \lambda S_0(\lambda) e^{\lambda(S_0(\lambda)x + m_0(\lambda))} \bar{F}_Y(S_0(\lambda)x + m_0(\lambda))/C_0(\lambda), \qquad \lambda > 0.$$

The moment-generating function of g_λ is given by:

$$
\begin{aligned}
\varphi_\lambda(t) &= \int_{-\infty}^{+\infty} e^{tx} g_\lambda(x) dx \\
&= \int_{-\infty}^{+\infty} \lambda S_0(\lambda) e^{(t+\lambda S_0(\lambda))x + \lambda m_0(\lambda)} \bar{F}_Y(S_0(\lambda)x + m_0(\lambda)) dx/C_0(\lambda) \\
&= \frac{\lambda}{\lambda + t/S_0(\lambda)} \exp\{-tm_0(\lambda)/S_0(\lambda)\} \frac{C_0(\lambda + t/S_0(\lambda))}{C_0(\lambda)}.
\end{aligned}
$$

Using properties (15.8), (15.9) and (15.10), we have:

$$
\begin{aligned}
\varphi_\lambda(t) &= \exp\left\{ \ln \lambda - \ln\left(\lambda + \frac{t}{S_0(\lambda)}\right) - t\frac{m_0(\lambda)}{S_0(\lambda)} + \ln C_0(\lambda + t/S_0(\lambda)) \right. \\
&\qquad \left. - \ln C_0(\lambda) \right\} \\
&= \exp\left\{ \frac{t^2}{2} + O(1/\lambda) \right\} \\
&\xrightarrow[\lambda \to \infty]{} \exp\left\{ \frac{t^2}{2} \right\}.
\end{aligned}
$$

It follows that:

$$G_\lambda(x) \xrightarrow{d} H(x).$$

By the dominated convergence theorem and the inversion formula for characteristic functions, we have:

$$g_\lambda(x) = \frac{1}{2\pi} \int_{-\infty}^{+\infty} e^{-itx} \varphi_\lambda(it) dt$$

$$\underset{\lambda\to\infty}{\to} \frac{1}{2\pi} \int_{-\infty}^{+\infty} e^{-itx} \exp\left\{\frac{t^2}{2}\right\} dt = \frac{1}{\sqrt{2\pi}} \exp\left\{-\frac{x^2}{2}\right\}.$$

Choosing $x = 0$, we get:

$$\bar{F}_Y(m_0(\lambda)) \underset{\lambda\to\infty}{\sim} \frac{\exp(-\lambda m_0(\lambda))C_0(\lambda)}{\sqrt{2\pi}\lambda S_0(\lambda)},$$

then, by making the substitution $\lambda \to m_0^{-1}(\lambda)$, we obtain (15.12).

Invoking the equivalents, we establish the theorem.

The following lemma will be very useful for ulterior proofs.

Lemma 15.6.1 *Let* $]a, b[$ *be an interval on* $\mathbb{R}$, *bounded or not. Let* $\psi :]a, b[\to \mathbb{R}$ *be a* C^2-*function, with an unique extremum in* c. *Moreover, we suppose that* $\psi''(c) < 0$. *Let* $\varphi :]a, b[\to \mathbb{R}$ *be a continuous and strictly positive function, such that* $\int_a^b \varphi(x)e^{\psi(x)} dx$ *exists, then:*

$$\int_a^b \varphi(x)e^{\lambda\psi(x)} dx \underset{\lambda\to\infty}{\sim} \sqrt{\frac{2\pi}{-\lambda\psi''(c)}} \varphi(c)e^{\lambda\psi(c)}.$$

Proof of lemma 15.6.1. Firstly, note that the integrals $\int_a^b \varphi(x)e^{\lambda\psi(x)} dx$, $\lambda \in \mathbb{R}^+$, exist, since:

$$\int_a^b \varphi(x)e^{\lambda\psi(x)} dx \leq e^{(\lambda-1)\psi(c)} \int_a^b \varphi(x)e^{\psi(x)} dx.$$

We look for an equivalent of the integral $\int_a^b \varphi(x)e^{\lambda\psi(x)} dx$, when $\lambda \to \infty$.

Step 1:

The function ψ has the following development around c:

$$\psi(x) = \psi(c) + \frac{(x-c)^2}{2}\psi''(c) + o(x-c)^2.$$

Let $\eta \in]0, 1[$. There exists $\delta > 0$ such that if $0 \leq x - c < \delta$, then:

$$(1 - \eta)\varphi(c) \leq \varphi(x) \leq (1 + \eta)\varphi(c)$$

$$\frac{1}{2}(x - c)^2\psi''(c)(1 + \eta) \leq \psi(x) - \psi(c) \leq \frac{1}{2}(x - c)^2\psi''(c)(1 - \eta),$$

hence $\forall \lambda \in \mathbb{R}^+$:

$$(1-\eta)\varphi(c)\int_c^{c+\delta} e^{\frac{1}{2}\lambda(x-c)^2\psi''(c)(1+\eta)} e^{\lambda\psi(c)}\,dx \le \int_c^{c+\delta} \varphi(x)e^{\lambda\psi(x)}\,dx,$$

and

$$\int_c^{c+\delta} \varphi(x)e^{\lambda\psi(x)}\,dx \le (1+\eta)\varphi(c)\int_c^{c+\delta} e^{\frac{1}{2}\lambda(x-c)^2\psi''(c)(1-\eta)} e^{\lambda\psi(c)}\,dx.$$

After a change of variable, the integral is undervalued by:

$$\sqrt{\frac{2}{-\lambda\psi''(c)(1+\eta)}}(1-\eta)\varphi(c)e^{\lambda\psi(c)}\int_0^{\sqrt{\frac{-\lambda\psi''(c)(1+\eta)}{2}}\delta} e^{-u^2}\,du.$$

Moreover, we know that:

$$\int_0^{\sqrt{\frac{-\lambda\psi''(c)(1+\eta)}{2}}\delta} e^{-u^2}\,du \underset{\lambda\to\infty}{\to} \frac{1}{2}\sqrt{\pi}.$$

Then, there exists λ_0, such that $\forall \lambda > \lambda_0$:

$$(1-\eta)\varphi(c)\int_c^{c+\delta} e^{\frac{1}{2}\lambda(x-c)^2\psi''(c)(1+\eta)} e^{\lambda\psi(c)}\,dx$$
$$\ge (1-\eta)\sqrt{\frac{\pi}{-2\lambda\psi''(c)(1+\eta)}}(1-\eta)\varphi(c)e^{\lambda\psi(c)},$$
$$(1+\eta)\varphi(c)\int_c^{c+\delta} e^{\frac{1}{2}\lambda(x-c)^2\psi''(c)(1-\eta)} e^{\lambda\psi(c)}\,dx$$
$$\le (1+\eta)\sqrt{\frac{\pi}{-2\lambda\psi''(c)(1-\eta)}}(1+\eta)\varphi(c)e^{\lambda\psi(c)}.$$

We deduce that for $\lambda > \lambda_0$:

$$\frac{(1-\eta)^2}{\sqrt{(1+\eta)}}\sqrt{\frac{\pi}{-2\lambda\psi''(c)}}\varphi(c)e^{\lambda\psi(c)} \le \int_c^{c+\delta} \varphi(x)e^{\lambda\psi(x)}\,dx,$$

and:

$$\int_c^{c+\delta} \varphi(x)e^{\lambda\psi(x)}\,dx \le \frac{(1+\eta)^2}{\sqrt{(1-\eta)}}\sqrt{\frac{\pi}{-2\lambda\psi''(c)}}\varphi(c)e^{\lambda\psi(c)}.$$

Step 2:

Let $\varepsilon > 0$. Choose η in $]0, 1[$, such that :

$$\frac{(1 - \eta)^2}{\sqrt{(1 + \eta)}} > 1 - \varepsilon \qquad \text{and} \qquad \frac{(1 + \eta)^2}{\sqrt{(1 - \eta)}} < 1 + \varepsilon.$$

Choose also δ and λ_0.

Step 3:

We must now check that the second part is negligible. By the assumptions, ψ is strictly decreasing at the right of c, then $\forall x > c + \delta$:

$$\psi(x) - \psi(c) \leq \psi(c + \delta) - \psi(c) = -\mu,$$

where μ is strictly positive. We deduce that for $\lambda \geq 1$:

$$\lambda\psi(x) \leq (\lambda - 1)\psi(c) - (\lambda - 1)\mu + \psi(x),$$

and:

$$0 \leq \int_{c+\delta}^{b} \varphi(x) e^{\lambda\psi(x)} \, dx \leq e^{(\lambda-1)\psi(c)-(\lambda-1)\mu} \int_{c+\delta}^{b} \varphi(x) e^{\psi(x)} \, dx.$$

Remark that $e^{-(t-1)\mu} = o\left(\frac{1}{\sqrt{t}}\right)$, and there exists λ_1, such that $\forall \lambda > \lambda_1$ then:

$$\int_{c+\delta}^{b} \varphi(x) e^{\lambda\psi(x)} \, dx < \varepsilon \sqrt{\frac{\pi}{-2\lambda\psi''(c)}} \varphi(c) e^{\lambda\psi(c)}.$$

At least, we have for any $\lambda > \max(\lambda_0, \lambda_1)$:

$$(1 - \varepsilon)\sqrt{\frac{\pi}{-2\lambda\psi''(c)}} \varphi(c) e^{\lambda\psi(c)} < \int_{c}^{b} \varphi(x) e^{\lambda\psi(x)} \, dx$$
$$< (1 + 2\varepsilon)\sqrt{\frac{\pi}{-2\lambda\psi''(c)}} \varphi(c) e^{\lambda\psi(c)}.$$

Step 4:

The same method used on $]a, c[$ give the same results.

Finally, we obtain:

$$\int_{a}^{b} \varphi(x) e^{\lambda\psi(x)} \, dx \underset{\lambda \to \infty}{\sim} \sqrt{\frac{2\pi}{-\lambda\psi''(c)}} \varphi(c) e^{\lambda\psi(c)}.$$

Proposition 15.6.1 *Let f_0 be the stationary density in (15.1) when $a = 0$. We have $q_0(\lambda) = \ln E(\exp\{\lambda \ln X^2\}) = \ln\left[2\int_0^\infty \exp\{\lambda \ln x^2\} f_0(x)dx\right]$, then:*

$$q_0(\lambda) = \frac{\lambda \ln \lambda}{(1-\alpha)} + \frac{\left[\ln(2b\alpha^{\frac{\alpha}{1-\alpha}}) - 1\right]}{(1-\alpha)}\lambda - \frac{\ln 2}{2\ln\alpha}\ln\lambda + m + o(1)$$

if and only if

$$f_0(x) \underset{x\to\infty}{\sim} \frac{D}{x^d}\exp(-ex^c),$$

with:

$$c = 2(1-\alpha), \qquad e = \frac{\alpha^{-\frac{\alpha}{1-\alpha}}}{2b(1-\alpha)},$$

$$d = \alpha + \frac{(1-\alpha)}{\ln(\alpha)}\ln 2, \qquad m = \ln\left[2D\left(\frac{2}{ec}\right)^{(1-d)/c}\left(\frac{\pi}{c}\right)^{1/2}\right].$$

Proof of proposition 15.6.1. The theorem 15.3.1. gives the first implication.

Reciprocally, we note $f_0(x) = D|x|^{-d}\exp(-e|x|^c)h(|x|)$. As f_0 is equivalent to $Dx^{-d}\exp(-ex^c)$ and integrable, the function h has the following properties: $\lim_{x\to\infty} h(x) = 1$, and on any intervals $[0, B]$ with $B > 0$, there exist two constants $A_B > 0$ and $\beta_B \le 1$ such that for each $x \in [0, B]$: $h(x) < A_B x^{d-\beta_B}$. In particular, $\forall x \in \mathbb{R}^+$, we have $h(x) \le A_1 x^{d-\beta_1} + C$ where $C = \sup\{h(x), x > 1\}$.

By using lemma 15.6.1, we would like to obtain an equivalent of:

$$q_0(\lambda) = \ln E\left(\exp\{\lambda \ln X^2\}\right) = \ln\left[2\int_0^\infty \exp\{\lambda \ln x^2\} f_0(x)dx\right],$$

We have:

$$\exp\{\lambda \ln x^2\} f_0(x) = \frac{D}{x^d}\exp\{2\lambda \ln x - ex^c\} h(x).$$

The function $x \to 2\lambda \ln x - ex^c$ reaches its maximum in $x = \Omega\lambda^{1/c}$, with $\Omega = \left(\frac{2}{ec}\right)^{1/c}$. We do the change of variable: $x = \Omega u \lambda^{1/c}$, and we obtain:

$$Dx^{-d}\exp\{2\lambda \ln x - ex^c\}$$
$$= \exp\left\{2\lambda \ln\left(\Omega\lambda^{1/c}\right)\right\} D\lambda^{-d/c} u^{-d}\Omega^{-d}\exp\{2\lambda(\ln u - u^c/c)\}.$$

The function: $u \to \ln u - u^c/c$ reaches its maximum in: $u_m = 1$. After the change of variable, $h(x)$ becomes $h(\Omega u \lambda^{1/c})$. We analyze its behavior in infinity and around 0:

(i) behavior of $h(\Omega u \lambda^{1/c})$ in infinity: let $\varepsilon > 0$ and $u_0 > 0$ be fixed, then there exists λ_{u_0} such that $\forall u > u_0$ and $\forall \lambda > \lambda_{u_0}$:

$$\left| h(\Omega u \lambda^{1/c}) - 1 \right| < \varepsilon.$$

(ii) behavior of $h(\Omega u \lambda^{1/c})$ around 0: for $\delta > 0$, we have:

$$\int_0^{u_m - \delta} h(\Omega u \lambda^{1/c}) u^{-d} \exp\{2\lambda (\ln u - u^c/c)\}$$
$$< \int_0^{u_m - \delta} \left(A_1 \Omega^{d-\beta_1} \lambda^{(d-\beta_1)/c} u^{-\beta_1} + C u^{-d} \right) \exp\{2\lambda (\ln u - u^c/c)\} \, du,$$

We can use lemma 15.6.1 (by taking account of (i) and (ii)), with the functions:

$$\varphi(u) \;\; = \;\; u^{-d},$$
$$\psi(u) \;\; = \;\; 2(\ln u - u^c/c).$$

Remember that:

$$q_0(\lambda) = \ln \left[2D \exp\left\{ 2\lambda \ln \left(\Omega \lambda^{1/c} \right) \right\} \lambda^{(1-d)/c} \Omega^{1-d} \right.$$
$$\left. \int_0^\infty h(\Omega u \lambda^{1/c}) \varphi(u) \exp\{\lambda \psi(u)\} \, du \right].$$

We have then $a = 0$, $b = \infty$, and $c = 1$. The steps 1, 2 and 3 are the same by taking account of the remark (i).

For the step 4, we obtain that the integral is undervalued by (remark (ii)):

$$\int_0^{1-\delta} h(\Omega u \lambda^{1/c}) \varphi(u) \exp\{2\lambda \psi(u)\} \, du$$
$$< e^{(\lambda-1)\psi(1) - (\lambda-1)\mu}$$
$$\int_0^1 \left(A_1 \Omega^{d-\beta_1} \lambda^{(d-\beta_1)/c} u^{-\beta_1} + C u^{-d} \right) \exp\{2(\ln u - u^c/c)\} \, du.$$

But, $e^{-\lambda\mu} = o\left(\lambda^{(\beta_1 - d)/c - 1/2} \vee \lambda^{-1/2} \right)$, and we can conclude in the same way.

Finally, we obtain that:

$$q_0(\lambda) = \frac{\lambda \ln \lambda}{(1-\alpha)} + \frac{\left[\ln(2b\alpha^{\frac{\alpha}{1-\alpha}})-1\right]}{(1-\alpha)}\lambda - \frac{\ln 2}{2\ln \alpha}\ln \lambda + m + o(1).$$

Proof of theorem 15.3.2. We note:

$$\begin{aligned}
q(\lambda) &= \ln E(\exp\{\lambda \ln X^2\}) = \ln\left[2\int_0^\infty \exp\{\lambda \ln x^2\} f(x)dx\right]\\
p_a(\alpha\lambda) &= \ln E\left(\exp\{(\alpha\lambda)\ln(a/b + X_t^{2\alpha})/\alpha\}\right)\\
&= \ln\left[2\int_0^\infty \exp\{(\alpha\lambda)\ln(a/b + x^{2\alpha})/\alpha\} f(x)dx\right].
\end{aligned}$$

The following are equivalent:

$$f(x) \underset{x\to\infty}{\sim} f_0(x) \Leftrightarrow q(\lambda) = q_0(\lambda) + o(1) \qquad \text{(Proposition 15.6.1)}$$

But,

$$\begin{aligned}
q_0(\lambda) &= q_0(\alpha\lambda) + \lambda \ln \lambda + \lambda(\ln 2b - 1) + \ln 2/2 + o(1)\\
q(\lambda) &= p_a(\alpha\lambda) + \lambda \ln \lambda + \lambda(\ln 2b - 1) + \ln 2/2 + o(1)
\end{aligned}$$

and then,

$$\begin{aligned}
&\quad f(x) \underset{x\to\infty}{\sim} f_0(x)\\
\Leftrightarrow &\quad p_a(\alpha\lambda) - q_0(\alpha\lambda) = o(1)\\
\Leftrightarrow &\quad p_a(\alpha\lambda) - q(\alpha\lambda) = o(1)\\
\Leftrightarrow &\quad \frac{\int_0^\infty \exp\{\lambda \ln(a/b + x^{2\alpha})\} f(x)dx}{\int_0^\infty \exp\{\lambda \ln x^{2\alpha}\} f(x)dx} \underset{\lambda\to\infty}{\to} 1\\
\Leftrightarrow &\quad \frac{\int_0^\infty \exp\{\lambda \ln x^{2\alpha}\}(\exp\{\lambda \ln(1 + a/bx^{2\alpha})\} - 1) f(x)dx}{\int_0^\infty \exp\{\lambda \ln x^{2\alpha}\} f(x)dx} \underset{\lambda\to\infty}{\to} 0\\
\Leftrightarrow &\quad \frac{\int_0^\infty \exp\{\lambda \ln x^{2\alpha}\}(\exp\{\lambda \ln(1 + a/bx^{2\alpha})\} - 1) k(x)f_0(x)dx}{\int_0^\infty \exp\{\lambda \ln x^{2\alpha}\} k(x)f_0(x)dx} \underset{\lambda\to\infty}{\to} 0,
\end{aligned}$$

where k is a function such that $\lim_{x\to\infty} k(x) = 1$.

To obtain an equivalent of the last one, we cut the integral in two parts:

$$\int_0^\infty = \int_0^{x_\lambda} + \int_{x_\lambda}^\infty.$$

The difficulty is to find a good speed for x_λ. To do it, let $x_\lambda = \lambda^v$.

For $x > x_\lambda$, we have:

$$\left| \exp\left\{ \lambda \ln(1 + a/bx^{2\alpha}) \right\} - 1 \right| < \exp\left\{ \frac{a\lambda^{1-2\alpha v}}{b} \right\} - 1.$$

We suppose that we have the following condition:

$$1 - 2\alpha v < 0 \qquad \text{or} \qquad v > 1/2\alpha. \tag{15.13}$$

Let $\varepsilon > 0$, there exists λ_0 such that for each $\lambda > \lambda_0$:

$$\frac{\int_{x_\lambda}^\infty \exp\left\{ \lambda \ln x^{2\alpha} \right\} \left(\exp\left\{ \lambda \ln(1 + a/bx^{2\alpha}) \right\} - 1 \right) k(x) f_0(x) dx}{\int_0^\infty \exp\left\{ \lambda \ln x^{2\alpha} \right\} k(x) f_0(x) dx} < \frac{\varepsilon}{2}.$$

We must now prove that the second part also tends to 0. To do it, we do the same operations as in the previous proof. We do the change of variable:

$$x = \Omega u(\alpha\lambda)^{1/c}.$$

The boundaries of the integral are modified:

$$\int_0^{x_\lambda} \rightarrow \int_0^{\Omega^{-1}\alpha^{-1/c}\lambda^{v-1/c}},$$

and if:

$$v - \frac{1}{c} < 0 \qquad \text{or} \qquad v < \frac{1}{2(1-\alpha)}, \tag{15.14}$$

then $\lambda^{v-1/c}$ tends to 0 when λ tends to infinity. If we use the end of the proof lemma 15.6.1, we can deduce that:

$$\int_0^{x_\lambda} \exp\left\{ \lambda \ln x^{2\alpha} \right\} k(x) f_0(x) dx = o\left(\int_0^\infty \exp\left\{ \lambda \ln x^{2\alpha} \right\} k(x) f_0(x) dx \right).$$

We want to prove now that also:

$$\int_0^{x_\lambda} \exp\left\{ \lambda \ln x^{2\alpha} \right\} \exp\left\{ \lambda \ln(1 + a/bx^{2\alpha}) \right\} k(x) f_0(x) dx$$
$$= o\left(\int_0^\infty \exp\left\{ \lambda \ln x^{2\alpha} \right\} k(x) f_0(x) dx \right).$$

To do it, we do the same change of variable. Respect to the previous one, we have an additional part:

$$\exp\left\{ \lambda \ln(1 + a/bx^{2\alpha}) \right\} = \exp\left\{ \lambda \ln\left(1 + a/\left(b(\Omega u(\alpha\lambda)^{1/c})^{2\alpha} \right) \right) \right\}.$$

Furthermore:

$$\int_0^{x_\lambda} \exp\left\{\lambda \ln x^{2\alpha}\right\} \exp\left\{\lambda \ln(1 + a/bx^{2\alpha})\right\} k(x) f_0(x) dx$$

$$= \frac{\exp\left\{2\alpha\lambda \ln\left(\Omega\lambda^{1/c}\right)\right\}}{\lambda^{(d-1)/c}} \times$$

$$\dots \int_0^{\Omega^{-1}a^{-1/c}\lambda^{v-1/c}} \frac{Dk(\Omega u\lambda^{1/c})}{u^d} \Omega^{1-d}$$

$$\exp\left\{\lambda\left(\ln\left(u^{2\alpha} + \frac{a}{b\Omega^{2\alpha}(\alpha\lambda)^{2\alpha/c}}\right) - 2\alpha u^c/c\right)\right\} du.$$

We note:

$$\psi_\lambda(u) = \alpha^{-1}\left(\ln\left(u^{2\alpha} + \frac{a}{b\Omega^{2\alpha}(\alpha\lambda)^{2\alpha/c}}\right) - 2\alpha u^c/c\right),$$

$$c_\lambda = \arg\max_{u\in[0,\infty[} \psi_\lambda(u).$$

We have the following properties:

- $\lim_{\lambda\to\infty} c_\lambda = 1$, moreover:

$$c_\lambda = 1 - \frac{a}{bc\Omega^{2\alpha}} \frac{1}{(\alpha\lambda)^{2\alpha/c}} + o\left(\frac{1}{\lambda^{2\alpha/c}}\right).$$

- on any interval $[U, \infty[$ with $U > 0$, there is uniform convergence of ψ_λ to ψ.

Then we deduce that there exists λ_1 such that $\forall \lambda > \lambda_1$:

$$\psi_\lambda(1 - \delta) - \psi(1) < 0.$$

And, $\forall u < 1 - \delta$:

$$\psi_\lambda(u) - \psi_\lambda(1 - \delta) < 0,$$

the end of the proof of lemma 15.6.1 (step 4) is valid and we can conclude in the same way.

If $\alpha > 1/2$, one can find a v which satisfies the constraints (15.13) and (15.14), and then:

$$q(\alpha\lambda) - p_a(\alpha\lambda) = o(1).$$

On the contrary, if we suppose that $\alpha \leq \frac{1}{2}$, then there exists λ_m such that $u_m - \delta < u < u_m + \delta$ and $\forall \lambda > \lambda_m$:

$$\lambda \ln(1 + a/b(\Omega u \lambda^{1/c})^{2\alpha}) > \frac{a\lambda^{\frac{1-2\alpha}{(1-\alpha)}}}{2b(u_m + \delta)^{2\alpha}\Omega^{2\alpha}},$$

and it follows that:

$$q(\alpha\lambda) - p_a(\alpha\lambda) \not\to 0.$$

At least, we conclude that:

$$f(x) \underset{x \to \infty}{\sim} f_0(x) \Leftrightarrow 1/2 < \alpha < 1.$$

An application of the Hospital rule yields the shape of $\bar{F}$.

Proof of theorem 15.4.3. We define the sequences:

$$
\begin{aligned}
d_n^Z &= \left(\frac{\ln n}{e}\right)^{2/c} + \frac{2}{c}\left(\frac{\ln n}{e}\right)^{2/c-1}\left(-\frac{f}{ec}\ln\left(\frac{\ln n}{e}\right) + \frac{\ln 2D}{e}\right), \\
c_n^Z &= \frac{2}{ec}\left(\frac{\ln n}{e}\right)^{2/c-1},
\end{aligned}
$$

such that if:

$$u_n^Z(\tau) = c_n^Z(-\ln \tau) + d_n^Z,$$

then:

$$n\left(1 - F_Z(u_n^Z(\tau))\right) \underset{n \to \infty}{\to} \tau.$$

Since $(X_t)_{t \in \mathbb{N}}$ is geometrically strong mixing, $(Z_t)_{t \in \mathbb{N}}$ is geometrically strong mixing too, and then the condition $D(u_n^Z(\tau))$ holds for (Z_t) with an upper bound $\alpha_{n,l_n} \leq Const\rho^{l_n}$ such that $\rho < 1$.

We introduce now an auxiliary process $(Y_t)_{t \in \mathbb{N}}$:

$$Y_t = \ln X_t^2 = \ln Z_t.$$

We have then:

$$Y_t = \alpha Y_{t-1} + \ln(b\epsilon_t^2) + \ln\left(1 + \frac{a}{b}e^{-\alpha Y_{t-1}}\right).$$

We note:

$$U_t = \ln(b\epsilon_t^2) \qquad \text{and} \qquad P_t = \ln\left(1 + \frac{a}{b}e^{-aY_{t-1}}\right),$$

and we define the autoregressive process of order 1 $(M_t)_{t\in\mathbb{N}}$, in the following way:

$$\begin{aligned} M_0 &= Y_0, \\ M_t &= \alpha M_{t-1} + U_t. \end{aligned}$$

We have then:

$$V_t = Y_t - M_t = \sum_{j=0}^{t-1} \alpha^j P_{t-j}.$$

Remark here that the random variables P_t and V_t are always positive and that:

$$n\left(1 - F_Y(u_n^Y(\tau))\right) \xrightarrow[n\to\infty]{} \tau, \qquad \text{with } u_n^Y(\tau) = \ln(u_n^Z(\tau)).$$

Let u be a threshold. We define $N_u = \inf\{j \geq 1 | Y_j \leq u\}$.

If we suppose that $Y_0 > u$, then we have for any $t \leq N_u$:

$$M_t \leq Y_t \leq M_t + \kappa(u), \qquad \text{a.s.,}$$

with:

$$\kappa(u) = \frac{a}{b(1-\alpha)}e^{-au}.$$

Now, we want to check the condition $D'(u_n^Z(\tau))$, i.e.:

$$\lim_{n\to\infty} n \sum_{j=1}^{p_n} P\{Z_0 > u_n^Z, Z_j > u_n^Z\} = 0.$$

Since $u_n^Z \to \infty$ when $n \to \infty$, we suppose that $u_n^Z > e^u$. We have:

$$n\sum_{j=1}^{p_n} P\{Z_0 > u_n^Z, Z_j > u_n^Z\} \leq n\sum_{j=1}^{p_n} P\{j > N_u, Z_0 > u_n^Z, Z_j > u_n^Z\}$$

$$+ n\sum_{j=1}^{p_n} P\{j \leq N_u, Z_0 > u_n^Z, Z_j > u_n^Z\}$$

$$\leq I_1 + I_2$$

To get an upper bound of I_1, we show first, as Borkovec (1999), that there exist constants $C > 0$ and $n_0 \in \mathbb{N}$ such that for any $n > n_0$, $z \in [0, e^u]$, $k \in \mathbb{N}^*$:

$$nP(Z_k > u_n^Z | Z_0 = z) \leq C.$$

Assume that it does not hold. Choose C and $N > 0$ arbitrary and $\eta > 0$ small. There exist $n > N$, $z \in [0, e^u]$ and $\delta(\eta) > 0$, such that for any $y \in]z - \delta, z + \delta[\cap [0, e^u]$, we have:

$$nP(Z_k > u_n^Z | Z_0 = y) > C - \eta.$$

But, we have also:

$$\lim_{n \to \infty} n \left(1 - F_Z(u_n^Z(\tau)) \right) = \tau,$$

for any τ as small as you want, and:

$$
\begin{aligned}
n \left(1 - F_Z(u_n^Z(\tau)) \right) &= \int_0^\infty nP(Z_k > u_n^Z | Z_0 = y) dF_Z(y) \\
&\geq \int_{]z-\delta, z+\delta[\cap [0,e^u]} nP(Z_k > u_n^Z | Z_0 = y) dF_Z(y) \\
&> (C - \eta) P(Z_0 \in]z - \delta, z + \delta[\cap [0, e^u]) \\
&\geq (C - \eta) D
\end{aligned}
$$

where $D = \inf_{z \in [0,e^u]} (F_Z(z + \delta) - F_Z(z)) > 0$ because F_Z is continuous. Since $C > 0$ is arbitrary, there is a contradiction.

Now, we have:

$$
\begin{aligned}
I_1 &\leq \sum_{j=1}^{p_n} \sum_{l=1}^{j-1} nP \left\{ N_u = l, Z_0 > u_n^Z, Z_j > u_n^Z \right\} \\
&\leq \sum_{j=1}^{p_n} \sum_{l=1}^{j-1} nP \left\{ Z_0 > u_n^Z, Z_l < e^u, Z_j > u_n^Z \right\}
\end{aligned}
$$

Let $C_n =]u_n^Z, \infty[$ and $D = [0, e^u]$. We note : $X_1 = Z_0$, $X_2 = Z_l$,

$X_3 = Z_j$. We have :

$$P\left\{Z_0 > u_n^Z, Z_l < e^u, Z_j > u_n^Z\right\}$$

$$= \int_{\mathbb{R}_+^3} \mathbb{I}_{\{x_1 \in C_n, x_2 \in D, x_3 \in C_n\}} P_{X_1, X_2, X_3}(x_1, x_2, x_3) dx_1 dx_2 dx_3$$

$$= \int_{\mathbb{R}_+^3} \mathbb{I}_{\{x_1 \in C_n, x_2 \in D, x_3 \in C_n\}} P_{X_3|X_2=x_2, X_1=x_1}(x_3) dx_3 P_{X_2|X_1=x_1}(x_2) dx_2$$
$$P_{X_1}(x_1) dx_1$$

$$= \int_{\mathbb{R}_+^3} \mathbb{I}_{\{x_1 \in C_n, x_2 \in D, x_3 \in C_n\}} P_{X_3|X_2=x_2}(x_3) dx_3 P_{X_2|X_1=x_1}(x_2) dx_2 P_{X_1}(x_1) dx_1$$

$$= \int_{\mathbb{R}_+^2} P(X_3 \in C_n | X_2 = x_2) \mathbb{I}_{\{x_1 \in C_n, x_2 \in D\}} P_{X_2|X_1=x_1}(x_2) dx_2 P_{X_1}(x_1) dx_1$$

$$\leq \frac{C}{n} \int_{\mathbb{R}_+^2} \mathbb{I}_{\{x_1 \in C_n, x_2 \in D\}} P_{X_2|X_1=x_1}(x_2) dx_2 P_{X_1}(x_1) dx_1$$

$$\leq \frac{C}{n} P(X_1 \in C_n, X_2 \in D) \leq \frac{C}{n} P(X_1 \in C_n) \underset{n \to \infty}{\sim} \frac{C\tau}{n^2}.$$

At least, we have:

$$I_1 \leq \sum_{j=1}^{p_n} j \frac{2C\tau}{n} \leq C\tau \frac{p_n^2}{n}$$

Furthermore, we have:

$$I_2 \leq n \sum_{j=1}^{p_n} P\left\{M_0 > u_n^Y, M_j > u_n^Y - \kappa(u)\right\}.$$

Moreover, $M_k = U_k + ... + \alpha^{k-1} U_1 + \alpha^k M_0$ and then:

$$\left\{M_k > u_n^Y - \kappa(u)\right\} \subset \left\{U_k + ... + \alpha^{k-1} U_1 > (u_n^Y - \kappa(u))/2\right\}$$
$$\cup \left\{\alpha^k M_0 > (u_n^Y - \kappa(u))/2\right\},$$

hence:

$$P\left\{M_0 > u_n^Y, M_j > u_n^Y - \kappa(u)\right\}$$
$$\leq P\left\{M_0 > u_n^Y, \alpha^k M_0 > (u_n^Y - \kappa(u))/2\right\}$$
$$+ P\left\{M_0 > u_n^Y\right\} P\left\{U_k + ... + \alpha^{k-1} U_1 > (u_n^Y - \kappa(u))/2\right\}.$$

Note that $\alpha > 1/2$. We choose $\varepsilon > 0$ such that $1 + \varepsilon < 1/(2\alpha)$. It exists n_u such that for all $n > n_u$, we have :

$$\frac{(u_n^Y - \kappa(u))}{2\alpha^k} > \frac{(u_n^Y - \kappa(u))}{2\alpha} > (1 + \varepsilon) u_n^Y > u_n^Y.$$

We deduce that :

$$P\left\{M_0 > u_n^Y, \alpha^k M_0 > (u_n^Y - \kappa(u))/2\right\}$$
$$\leq P\left\{M_0 > (1+\varepsilon)u_n^Y\right\} = P\left\{Z_0 > (u_n^Z(\tau))^{(1+\varepsilon)}\right\}.$$

But, we have :

$$(u_n^Z(\tau))^{(1+\varepsilon)} = \left(\frac{\ln n}{e}\right)^{2(1+\varepsilon)/c} A_n^{(1+\varepsilon)}(\tau), \qquad \text{où } \lim_{n\to\infty} A_n(\tau) = 1,$$

and then :

$$P\left\{Z_0 > (u_n^Z(\tau))^{(1+\varepsilon)}\right\}$$
$$\leq \exp\left\{-(\ln n)^{(1+\varepsilon)} A_n^{c/2}(\tau) - f\ln\left((\ln n/e)^{2(1+\varepsilon)/c} A_n^{(1+\varepsilon)}(\tau)\right) + \ln(2D)\right\}$$
$$\leq \exp\left\{-2(\ln n)^{(1+\varepsilon)}\right\}$$

for any n big enough.

Furthermore, it is easy to see that the bigger k, the heavier the distribution tail of the random variable $U_k + ... + \alpha^{k-1}U_1$, by using lemma 1.6.1 for example. And in the same way as before, we have that for any n big enough:

$$P\left\{U_k + ... + \alpha^{k-1}U_1 > (u_n^Y - \kappa(u))/2\right\} \leq P\left\{Y_0 > (u_n^Y - \kappa(u))/2\right\}$$
$$\leq \exp\left\{-2(\ln n)^{1/2-\varepsilon}\right\}$$

It follows that:

$$I_2 \leq p_n\left(const\exp\left\{-2(\ln n)^{1+\varepsilon}\right\} + const\exp\left\{-2(\ln n)^{1/2-\varepsilon}\right\}\right)$$
$$I_2 \leq p_n const\exp\left\{-2(\ln n)^{1/2-\varepsilon}\right\}$$

where $const$ is a generic constant.

Finally, by choosing:

$$p_n = \left[\exp\left\{(\ln n)^{1/4}\right\}\right] \qquad \text{and} \qquad l_n = \left[\exp\left\{(\ln n)^{1/8}\right\}\right]$$

then all the conditions for $D'(u_n^Z(\tau))$ are verified, the statement follows: the extremal index of (Z_t), θ_Z, exists and is equal to one.

15.7   Conclusion

We observe quite different extremal behaviors depending on whether $\alpha = 1$ or $\alpha < 1$. In the first case, we observe Pareto-like tails and an extremal index which is strictly less than one. In the second case, for $\alpha > 1/2$, the tails are Weibull–like and the extremal index is equal to one.

APPENDIX

Appendix 1:

We define the functions:

$$g_1(\lambda) = \lambda \ln \lambda, \qquad g_2(\lambda) = \lambda, \qquad \text{and} \qquad g_3(\lambda) = \ln \lambda.$$

We have then:

$$\begin{aligned}
g_1(\lambda) - g_1(\alpha\lambda) &= (1-\alpha)\lambda \ln \lambda - \lambda\alpha \ln \alpha, \\
g_2(\lambda) - g_2(\alpha\lambda) &= (1-\alpha)\lambda, \\
g_3(\lambda) - g_3(\alpha\lambda) &= -\ln \alpha.
\end{aligned}$$

There exist three constants a_1, a_2 et a_3 such that:

$$a_1(g_1(\lambda) - g_1(\alpha\lambda)) + a_2(g_2(\lambda) - g_2(\alpha\lambda)) + a_3(g_3(\lambda) - g_3(\alpha\lambda))$$
$$= \lambda \ln \lambda + \lambda(\ln 2b - 1) + \ln 2/2,$$

which are given by:

$$a_1 = \frac{1}{(1-\alpha)}, \qquad a_2 = \frac{\left[\ln\left(2b\alpha^{\frac{\alpha}{1-\alpha}}\right) - 1\right]}{(1-\alpha)} \qquad \text{et} \qquad a_3 = -\frac{\ln 2}{2\ln \alpha}.$$

And then,

$$q_0(\lambda) - q_0(\alpha\lambda) = a_1(g_1(\lambda) - g_1(\alpha\lambda)) + a_2(g_2(\lambda) - g_2(\alpha\lambda))$$
$$+ a_3(g_3(\lambda) - g_3(\alpha\lambda)) + O(1/\lambda).$$

Remark now that if s is a function such that $s(\lambda) \underset{\lambda\to\infty}{\sim} c/\lambda$ with c different from 0, then:

$$s(\lambda) - s(\alpha\lambda) \underset{\lambda\to\infty}{\sim} \frac{c(1 - 1/\alpha)}{\lambda}$$

At least, we note C the set of continuous fonctions from $\mathbb{R}^+$ to $\mathbb{R}$, and we define the application $\Psi : C \to C$ such that $\Psi(f) = f_\alpha$ and $f_\alpha(\lambda) =$

$f(\lambda) - f(\alpha\lambda)$, $\forall \lambda \in \mathbb{R}^+$. The kernel of this linear application is the set of the constants.

With all these elements, we deduce that:

$$q_0(\lambda) = \frac{\lambda \ln \lambda}{(1-\alpha)} + \frac{\left[\ln\left(2b\alpha^{\frac{\alpha}{1-\alpha}}\right) - 1\right]}{(1-\alpha)}\lambda - \frac{\ln 2}{2\ln\alpha}\ln\lambda + m + O(1/\lambda).$$

Appendix 2: The process (Y_t) is defined by:

$$Y_t = [X_t] + U_t - 0.5.$$

We have the following inequalities:

$$-0.5 \le U_t - 0.5 \le 0.5 \qquad \text{and} \qquad X_t - 1 < [X_t] \le X_t,$$

and then,

$$X_t - 1.5 < Y_t \le X_t + 0.5.$$

We deduce that:

$$\frac{P(X_t > x + 1.5)}{P(X_t > x)} \le \frac{P(Y_t > x)}{P(X_t > x)} \le \frac{P(X_t > x - 0.5)}{P(X_t > x)}.$$

But, we have also:

$$\frac{P(X_t > x - 0.5)}{P(X_t > x)} \underset{x\to\infty}{\sim} \left(1 - \frac{1}{2x}\right)^f \exp\left\{\frac{ecx^{c-1}}{2}\right\} \underset{x\to\infty}{\to} 1,$$

$$\frac{P(X_t > x + 1,5)}{P(X_t > x)} \underset{x\to\infty}{\sim} \left(1 + \frac{3}{2x}\right)^f \exp\left\{-\frac{3ecx^{c-1}}{2}\right\} \underset{x\to\infty}{\to} 1,$$

and finally we obtain:

$$P(Y_t > x) \underset{x\to\infty}{\sim} P(X_t > x).$$

Bibliography

Beirlant, J., Broniatowski, M., Teugels, J. and Vynckkier, P. (1995). The mean residual life function at great age: Applications to tails estimation, *J. Stat. Plann. Inf* **45**: 21–48.

Bollerslev, T. (1986). Generalized autoregressive conditional heterosce-dasticity, *J. Econometrics* **31**: 307–327.

Bollerslev, T., Chou, R. and Kroner, K. (1992). Arch modelling in finance, *J. Econometrics* **52**: 5–59.

Borkovec, M. (1999). Extremal behavior of the autoregressive process with arch(1) errors, *Technical report*, University of München.

Breidt, F. and Davis, R. (1998). Extremes of stochastic volatility models, *Ann. Appl. Probab* **8**: 664–675.

Cont, R., Potters, M. and Bouchaud, J. (1997). Scaling in stock market data: Stable law and beyond, *in* D. et al. (ed.), *Scale Invariance and Beyond*, Springer, Berlin.

Davis, R., Mikosch, T. and Basrak, C. (1999). Sample acf of stochastic recurrence equations with application to garch, *Technical report*, University of Groningen.

Diebolt, J. and Guegan, D. (1991). Le modèle de série chronologique autorégressive β-arch, *Acad. Sci. Paris* **312**: 625–630.

Embrechts, P., C., K. and Mikosch, T. (1997). *Modelling Extremal Events*, Springer, Berlin.

Engle, R. (1982). Autoregressive conditional heteroscedasticity with estimates of the variance of united kingdom inflation, *Econometrica* **50**: 987–1008.

Gourieroux, C. (1997). *ARCH Models and Financial Applications*, Springer, Berlin.

Gourieroux, C., Jasiak, J. and Le Fol, G. (1999). Intraday market activity, *Journal of Financial Markets* **2**: 193–226.

Haan, L. D., Resnick, S., Rootzen, H. and Vries, C. D. (1989). Extremal behaviour of solutions to a stochastic difference equation with application to arch processes, *Stoch. Proc Appl.* **32**: 213–224.

Leadbetter, M., Lindgren, G. and Rootzen, H. (1983). *Extremes and Related Properties of Random Sequences and Processes*, Springer, Berlin.

Leadbetter, M. and Rootzen, H. (1988). Extremal theory for stochastic processes, *Ann. Probab.* **16**: 431–478.

Mandelbrot, B. (1963). The variation of certain speculative prices, *J. Business* **36**: 394–419.

Mikosch, T. and Starica, C. (1998). Limit theory for the sample autocorrelations and extremes of a garch(1,1) process, *Technical report*, University of Groningen.

Perfect, C. (1994). Extremal behavior of stationary markov chains with applications, *Ann. Appl. Probab* **4**: 529–548.

Resnick, S. (1987). *Extreme Values, Regular Variation, and Point Processes*, Springer, Berlin.

Robert, C. (2000). Mouvement extrêmes des séries financières haute fréquence, *Finance* .

Smith, R. and Weismann, L. (1994). Estimating the extremal index, *Journal of the Royal Statistical Society B* **56**: 515–528.

Lecture Notes in Statistics

For information about Volumes 1 to 73,
please contact Springer-Verlag

Vol. 74: P. Barone, A. Frigessi, M. Piccioni, Stochastic Models, Statistical Methods, and Algorithms in Image Analysis. vi, 258 pages, 1992.

Vol. 75: P. K. Goel, N.S. Iyengar (Eds.), Bayesian Analysis in Statistics and Econometrics. xi, 410 pages, 1992.

Vol. 76: L. Bondesson, Generalized Gamma Convolutions and Related Classes of Distributions and Densities. viii, 173 pages, 1992.

Vol. 77 E. Mammen, When Does Bootstrap Work? Asymptotic Results and Simulations. vi, 196 pages, 1992.

Vol. 78: L. Fahrmeir, B. Francis, R. Gilchrist, G. Tutz (Eds.), Advances in GLIM and Statistical Modelling: Proceedings of the GLIM92 Conference and the 7th International Workshop on Statistical Modelling, Munich, 13-17 July 1992 ix, 225 pages, 1992.

Vol. 79: N. Schmitz, Optimal Sequentially Planned Decision Procedures. xii, 209 pages, 1992.

Vol 80: M. Fligner, J. Verducci (Eds.), Probability Models and Statistical Analyses for Ranking Data. xxii, 306 pages, 1992.

Vol 81: P. Spirtes, C. Glymour, R. Scheines, Causation, Prediction, and Search. xxiii, 526 pages, 1993.

Vol. 82: A. Korostelev and A. Tsybakov, Minimax Theory of Image Reconstruction. xii, 268 pages, 1993.

Vol. 83: C. Gatsonis, J. Hodges, R. Kass, N. Singpurwalla (Editors), Case Studies in Bayesian Statistics. xii, 437 pages, 1993

Vol. 84: S. Yamada, Pivotal Measures in Statistical Experiments and Sufficiency. vii, 129 pages, 1994.

Vol. 85: P. Doukhan, Mixing: Properties and Examples. xi, 142 pages, 1994.

Vol 86: W. Vach, Logistic Regression with Missing Values in the Covariates. xi, 139 pages, 1994.

Vol 87: J. Muller, Lectures on Random Voronoi Tessellations.vii, 134 pages, 1994.

Vol. 88: J. E. Kolassa, Series Approximation Methods in Statistics. Second Edition, ix, 183 pages, 1997.

Vol. 89. P. Cheeseman, R.W. Oldford (Editors), Selecting Models From Data AI and Statistics IV xii, 487 pages, 1994.

Vol 90. A. Csenki, Dependability for Systems with a Partitioned State Space: Markov and Semi-Markov Theory and Computational Implementation x, 241 pages, 1994.

Vol. 91. J.D. Malley, Statistical Applications of Jordan Algebras. viii, 101 pages, 1994.

Vol. 92: M. Eerola, Probabilistic Causality in Longitudinal Studies. vii, 133 pages, 1994

Vol. 93: Bernard Van Cutsem (Editor), Classification and Dissimilarity Analysis. xiv, 238 pages, 1994.

Vol. 94: Jane F. Gentleman and G.A. Whitmore (Editors), Case Studies in Data Analysis. viii, 262 pages, 1994.

Vol. 95. Shelemyahu Zacks, Stochastic Visibility in Random Fields. x, 175 pages, 1994.

Vol. 96: Ibrahim Rahimov, Random Sums and Branching Stochastic Processes. viii, 195 pages, 1995.

Vol. 97: R. Szekli, Stochastic Ordering and Dependence in Applied Probability. viii, 194 pages, 1995.

Vol. 98: Philippe Barbe and Patrice Bertail, The Weighted Bootstrap. viii, 230 pages, 1995.

Vol. 99: C. C. Heyde (Editor), Branching Processes: Proceedings of the First World Congress. viii, 185 pages, 1995.

Vol. 100: Wlodzimierz Bryc, The Normal Distribution: Characterizations with Applications. viii, 139 pages, 1995

Vol. 101 H.H. Andersen, M.Højbjerre, D. Sørensen, P.S.Eriksen, Linear and Graphical Models. for the Multivariate Complex Normal Distribution. x, 184 pages, 1995.

Vol. 102: A.M Mathai, Serge B. Provost, Takesi Hayakawa, Bilinear Forms and Zonal Polynomials x, 378 pages, 1995

Vol. 103: Anestis Antoniadis and Georges Oppenheim (Editors), Wavelets and Statistics. vi, 411 pages, 1995.

Vol. 104: Gilg U.H. Seeber, Brian J. Francis, Reinhold Hatzinger, Gabriele Steckel-Berger (Editors), Statistical Modelling: 10th International Workshop, Innsbruck, July 10-14th 1995. x, 327 pages, 1995.

Vol. 105: Constantine Gatsonis, James S. Hodges, Robert E. Kass, Nozer D. Singpurwalla(Editors), Case Studies in Bayesian Statistics, Volume II. x, 354 pages, 1995

Vol. 106: Harald Niederreiter, Peter Jau-Shyong Shiue (Editors), Monte Carlo and Quasi-Monte Carlo Methods in Scientific Computing. xiv, 372 pages, 1995.

Vol. 107: Masafumi Akahira, Kei Takeuchi, Non-Regular Statistical Estimation. vii, 183 pages, 1995.

Vol. 108 Wesley L. Schaible (Editor), Indirect Estimators in U.S. Federal Programs viii, 195 pages, 1995.

Vol. 109. Helmut Rieder (Editor), Robust Statistics, Data Analysis, and Computer Intensive Methods. xiv, 427 pages, 1996.

Vol. 110 D. Bosq, Nonparametric Statistics for Stochastic Processes. xii, 169 pages, 1996.